Introducing
Geomorphology

Companion titles:

Introducing Geology – A Guide to the World of Rocks (Second Edition 2010)
Introducing Palaeontology – A Guide to Ancient Life (2010)
Introducing Volcanology ~ A Guide to Hot Rocks (2011)
Introducing Meteorology ~ A Guide to the Weather (forthcoming 2012)
Introducing Tectonics, Rock Structures and Mountain Belts (forthcoming 2012)
Introducing Oceanography (forthcoming 2012)

For further details of these and other Dunedin
Earth and Environmental Sciences titles see
www.dunedinacademicpress.co.uk

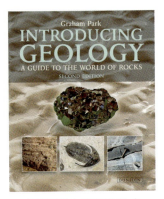

ISBN 978-1-906716-21-9

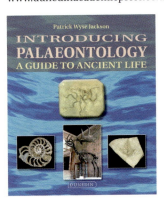

ISBN 978-1-906716-15-8

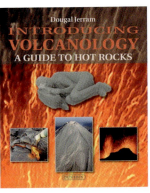

ISBN 978-1-906716-22-6

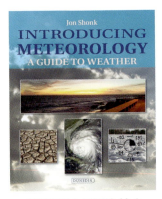

ISBN 978-1-780460-02-4

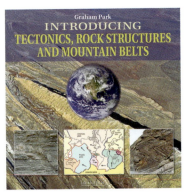

ISBN 978-1-906716-26-4

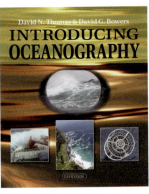

ISBN 978-1-780460-01-7

Introducing Geomorphology

A Guide to Landforms and Processes

Adrian Harvey

DUNEDIN

Published by
Dunedin Academic Press Ltd
Hudson House
8 Albany Street
Edinburgh EH1 3QB
Scotland

www.dunedinacademicpress.co.uk

ISBN 978-1906716-32-5

British Library Cataloguing in Publication data
A catalogue record for this book is available from the British Library

Design and layout by Makar Publishing Production, Edinburgh
Printed in Poland by Hussar Books

Contents

List of illustrations and tables

Preface

Many people enjoy scenery, and see landscapes as beautiful. Landscapes have inspired painters and photographers, even poets and composers of music. What is it about scenery that inspires people? For some it is clearly related to a scene with a 'natural' vegetation cover, for others it might relate to a mood created by the weather, or atmospheric conditions. For most, the physical landscape (mountains, hills, rocks, rivers, the sea) is the basis of 'scenery'. Geomorphology is the science that deals with landforms and physical landscapes. As a study it lies between the traditional disciplines of physical geography and geology, draws from both, and contributes to both.

The purpose of this book is to introduce the reader to the science of geomorphology. The book is not intended as a textbook; there are many of these at every level. I make no bibliographic references in the text, but in the end pages I make suggestions for further reading. In writing I have tried to present a broad and reasonably comprehensive view of the conceptual basis of the subject. I have tried to avoid mathematical treatments and to keep the level of physics and chemistry to a minimum.

Geomorphology inevitably involves a great range of spatial scales, from the global scale (continents and mountain systems) to the regional scale (individual mountain and hill ranges, river basins) to the local scale (conventional scenery: rivers, hillslopes, beaches, glaciers) and the micro-scale (weathering phenomena, sedimentary details). Partly related to spatial scales are timescales (geological time – millions of years; the timescales of the ice ages – half a million to tens of thousands of years; modern timescales – the last ten thousand years; timescales of individual events, e.g. landslides, floods – hours or days). The form of the earth's surface, at all scales, results from the interplay between two sets of forces, though the relative importance of each varies with scale. The two sets of forces are internal (essentially geologically driven) and external (essentially climatically driven).

This book is organised from the 'top down', initially introducing the concepts related to spatial and temporal scales and the two main drivers of landform evolution (internal and external forces). Then the bulk of the material is organised by spatial scale, dealing first with global and regional scales, then with local and (to some extent) micro-scales. There are two final short chapters, one dealing with the integration of timescales and landscape evolution, the other dealing with interactions between human society and geomorphology.

Note: all terms highlighted in **bold** are defined in the Glossary at the end of the book.

Acknowledgements

In the writing of this book I am grateful to Dunedin Academic Press for their support and encouragement; to the staff of the School of Environmental Sciences (formerly the Department of Geography) at the University of Liverpool, throughout my career, and particularly to Sandra Mather, head of the Cartographics section there, for her unstinting attention to detail in preparing the maps and diagrams. I thank my family for their encouragement, my son Michael for providing the background cover photo. With the exception of those images downloaded from Google Earth, the other photos, good or bad, are my own. Finally, I especially thank my wife, Karina, for her constructive criticisms of my writing style, but particularly for her patience and understanding throughout the preparation of this book.

1 Introduction to geomorphology

In this chapter I introduce some fundamental concepts related to spatial and temporal scales of landforms and some basic material on the primary forces driving landform evolution: internal (geological) and external (climatic) forces.

1.1 What do we mean by 'landforms'?

Geomorphology is the scientific study of the landforms of the surface of the Earth. These forms encompass a range of scales from that of, for example, the Earth's major plains, plateaux and mountain ranges to that of small-scale forms, such as a beach or a river bank. Landforms of various scales are nested within one another so that within a mountain range, for example, there are individual mountain ridges and valley systems; within valley systems there are valley-side slopes and river channels; and within river channels there are gravel and sand bars. Studying landforms and the processes that create them inevitably involves study over a range of temporal and spatial scales. The processes that create the landforms include the creation of the relief itself and its modification by erosion and deposition. The temporal scales range from the short-term timescales at which some erosion processes operate, to the longer-term timescales of Earth history. The spatial scales range from large scales related to the global distribution of the relief forms of the land surface and the sea floor to local scales of, for example, individual hillslopes or river channels.

As a science, geomorphology lies between the traditional disciplines of physical geography, the study of the natural environment, and geology, the study of the solid Earth. As the Earth's surface forms part of the natural environment, geomorphology interacts with the other sciences that deal with environmental systems: climatology, hydrology, pedology (soil science) and ecology. It also interacts with several subdisciplines of geology, especially with tectonics and structural geology in relation to deformation of the Earth's crust, with sedimentology in relation to the properties of sediments, the products of erosion at the Earth's surface, and with stratigraphy, the account of Earth history.

This interface between geology and environment influences the spatial and temporal scales relevant to the study of geomorphology. In the simplest terms, internally driven (geological) forces generally tend to operate over large spatial and long time scales. They create the gross or **available relief** of the Earth's surface by deformation of the Earth's crust. Externally driven forces, ultimately controlled by climate, modify this surface by erosion and deposition. These external processes can be described as the 'sediment cascade' and they create the more detailed landforms that form the heart of the study of geomorphology.

1.2 What we mean by spatial scales

At global and continental scales geomorphology deals with the major features of the surface of the Earth (e.g. continents, mountain systems). At a regional scale it deals with intermediate forms (e.g. individual mountain and hill ranges, river basins). At a local scale it deals with what could be described as individual features of conventional 'scenery' (e.g. rivers, hillslopes, beaches, glaciers), and at a micro-scale with the detail of the surface itself and its constituent materials (e.g. weathering phenomena, sedimentary details).

This book is organised around these themes, with the main chapters devoted to global scales, regional scales, with local and micro-scales treated together. Figure 1.1 shows how different features are apparent at different scales. On the Google Earth satellite image of western Canada (Figure 1A), features related to global/continental scales are most apparent. The NNW–SSE alignment of the major structures of the Rocky Mountains dominate the west of the image. These structures relate to the plate tectonics context (*see* below) of the North American continental plate, convergent with the oceanic Pacific plate. In contrast, the Canadian Prairies present a much more uniform land surface, formed over a much more stable zone of the Earth's crust. In addition to the geological features the vegetation cover reflects major continental-scale climatic contrasts. The grassland of the southern Prairies contrasts with the forest cover further north and with the forested zones within the Rocky Mountains. Coming down a scale to that of a major valley within the Rockies (Figure 1B), the NNW–SSE alignments of the major mountain ridges and of the Saskatchewan and Bow river valleys are still apparent, but the mountain slopes dominate the image. Coming right down to the local scale (Figure 1C), the characteristics of the braided river channel dominate the photograph.

1.3 What we mean by temporal scales

The three spatial scales represented by the views shown in Figure 1.1 all relate to different timescales. The geological processes that

Figure 1.1 Examples of scale in geomorphology, illustrated by the Canadian Rocky Mountains. **A.** Continental scale, represented by a satellite image – ©Google Earth image of the Rocky Mountains and adjacent areas of the Canadian Prairies. The timescale related to the development of the physical features depicted on the image is one of millions of years. Note the contrast between the structurally complex mountain chain and the structurally stable area to the east, where the plains are underlain by near-horizontal bedrock. Note also how the satellite image brings out the vegetation contrasts between the grasslands of the southern

Prairies and the forested areas further north and in the mountains. **B.** Regional (landscape) scale photograph looking south along the Bow Valley within the Rocky Mountains. At this scale the straight alignment of the main valley is clear, as is the mountain morphology to the west. The westerly dip of the sedimentary rocks forming the mountains is clear, as are the glacial erosional features on the mountains and the scree slopes at the base. The timescale related to the development of the landforms visible in the photo is one of tens of thousands of years. **C.** Detailed photograph of the bed of the braided river channel of the North Saskatchewan River within the Canadian Rockies. Note the milky colour of the water, due to suspended sediment, released by melting glaciers within the catchment of the river. Note also the gravel bars on the bed of the river. This river has only existed for the last 8000 years or so, and its detailed form is modified at least annually by snowmelt floods.

formed the structures of the Rockies have been operating over 30–50 million years, but their **tectonic** uplift to form a mountain chain has probably occurred only over the last 10 million years or so. In the second view we are looking at a landscape that was periodically covered by ice sheets during the last half million years, then by huge valley glaciers that deepened the main valleys. The small glaciers in the Rockies today are merely diminutive remnants of much bigger bodies of ice that melted only 8–10 thousand years ago. The third view shows a river that has only existed for maybe 8000 years, but its detailed morphology is modified following every flood generated by summer storms or the annual snowmelt flood in the spring.

To grasp this wide range of timescales we need to know a little about geological time, and especially about the particular timescales relating to approximately the last two million years. The Earth as a planet is about 4000 million years old. These vast ages have been determined on the basis of the decay rates of radioactive elements contained within the rocks of the crust. The later 15% of geological time (the Phanerozoic; *see* Figure 1.2A) is divided on the basis of fossil evidence into a system of eras and periods, related to the evolution of life forms on the Earth. The landforms of the surface of the Earth are young in relation to geological time. Most of the detailed form has developed only over the last 1.6 million years or so during the Quaternary (the Pleistocene and the Holocene; Figure 1.2B). The present (Alpine) system of mountain ranges dates largely from the mid-Cenozoic (*c.*25 million years) and the modern pattern of continents and ocean basins dates largely from the early Cenozoic (65 million years). Only in parts of Australia and Africa can Mesozoic landform patterns be recognised to any great extent. However, older mountain systems can be recognised in the structural patterns and to a certain extent in the relief of all the continents. It is with Quaternary (*see* Figure 1.2B) timescales (the last 1.6 million years) that the geomorphologist is most concerned.

The modern basis for subdivision of the Quaternary is climate. Over the last 1.6 million years there have been numerous climatic oscillations. Interglacial conditions (such as the present time) have alternated with global glacial periods, when lower global temperatures allowed large ice sheets to form over much of the northern continental areas, in addition to the more permanent ice sheets over Greenland and Antarctica. These oscillations are caused by cyclic variations in the Earth's orbital characteristics (**Milankovitch cycles**: named after the Yugoslav mathematician who discovered them). We shall deal with the effects in Chapter 2. Here we deal with the modern basis for Quaternary chronology. The chronology is based on the oxygen isotope record preserved in ocean-floor sediments and in the glacial ice of Greenland and Antarctica. The two isotopes of oxygen (O^{16} and O^{18}) have different temperature-related potentials for evaporation. Hence, atmospheric concentrations of each isotope are enhanced and seawater concentrations diminished, or vice versa, as the result of fluctuations in the global temperature, and particularly in the volume of water in the oceans, which reflect global glacial/interglacial cycles. Oxygen isotope ratios preserved within the calcium carbonate ($CaCO_3$) of foraminifera shells (small marine protozoans) or within ice crystals therefore vary with the global glacial/interglacial cycles. Figure 1.2B summarises the oxygen isotope

EON	ERA	PERIOD		DATE (Ma) approx
PHANEROZOIC	Cenozoic (Tertiary)	Quaternary	Holocene	0.01
			Pleistocene	2
		Neogene	Pliocene	6
			Miocene	25
		Palaeogene	Oligocene	40
			Eocene	65
	Mesozoic	Cretaceous		135
		Jurassic		200
		Triassic		240
	(Upper) Palaeozoic	Permian		280
		Carboniferous		370
		Devonian		415
	(Lower)	Silurian		445
		Ordovician		515
		Cambrian		590
(PRE-CAMBRIAN)				
PROTEROZOIC				2500
ARCHAEAN				4000

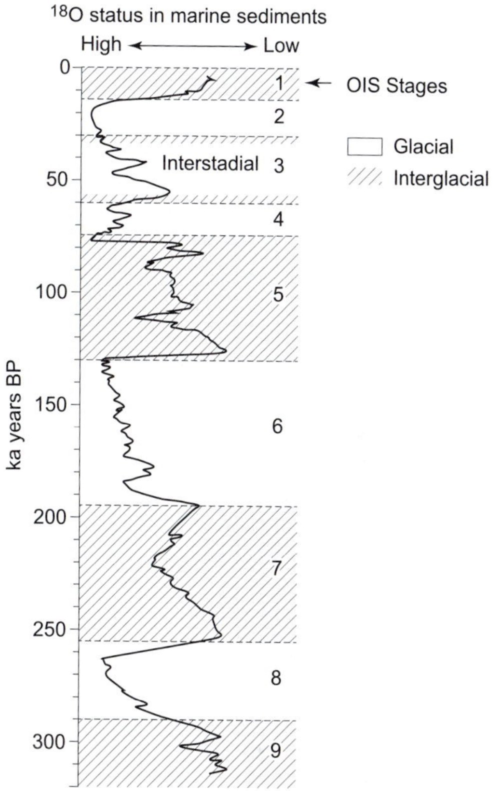

Figure 1.2 Geological timescales. **A.** *(Above)* Geological timescale. **B.** *(Right)* Timescale for the last 300,000 years of the Quaternary: based on the oxygen isotope record from foraminifera in marine sediments. The nomenclature for the stages here labelled OIS (Oxygen Isotope Stages) has recently been modified to MIS (Marine Isotope Stages).

record for last 300,000 years of the middle and late Quaternary derived from the marine record. The warm interglacial or mild interstadial phases are given odd numbers increasing in age from the modern Holocene, Oxygen Isotope Stage 1 (OIS 1). OIS 3 was an interstadial, not as warm as the Holocene, and the last major interglacial (OIS 5) occurred around 125,000–90,000 years ago. The intervening cold or glacial phases of the Pleistocene are numbered with even numbers increasing in age from the last glaciation (OIS 2). This and earlier glaciations during OIS 6 and OIS 8 were particularly important for geomorphology (*see* Section 1.4 below). The oxygen isotope-based chronology and notation has largely replaced the Alpine, Northern European or American regional terminology previously used to define glacial phases. The modern approach to Quaternary chronology is based on the global climatic sequence rather than on local and incomplete stratigraphic sequences.

The last 10,000 years (The Holocene: OIS 1) have been important in modifying the effects on the landscape of the environments of the last global glacial (OIS 2). They have also been important for another reason. For the latter half of the Holocene humans have had an increasing impact on the landscape. For the majority of this timespan the natural landscape was progressively modified by human settlement and the development of agriculture. The evidence for geomorphic change is intimately linked with that for vegetation change and with archaeological evidence for the development of human societies. Over the last 200 years or so the human impact on geomorphic systems has accelerated enormously, both indirectly through radical changes to the properties of the Earth's surface, especially through vegetation change, and more directly through engineering interventions in the sediment cascade. We will consider timescales of landform evolution in more detail in Chapter 5 and human impact on geomorphology in Chapter 6.

There is still another approach to timescales that is important in understanding geomorphic processes (Chapter 4). The effectiveness of individual geomorphic events (e.g. floods, landslides) tends to increase with their rarity. For example, a flood that occurs on average once every 100 years (ie. with a **recurrence interval** of 100 years) will bring about much more erosional change than a flood that occurs on average once every five years. However, because we might expect there to be 20 five-year floods for every 100-year flood, the overall and cumulative effects of the five-year floods may be greater. This concept, referred to as the **magnitude and frequency concept**, was developed in the 1960s in a classic paper by Gordon Wolman and John Miller (*see* Figure 1.3), where they demonstrated that in active landscapes the greatest cumulative amount of geomorphic work done (erosion, sediment transport, deposition) was carried out by events of moderate magnitude and frequency. Active landforms, especially in river systems, tend to adjust to such events by erosion and deposition. For example, river channels (*see* Chapter 4) tend to be maintained by a balance between erosion and deposition at a size related to such moderate events.

Two major elaborations of this concept were developed particularly in the concept of **geomorphic sensitivity** by Dennis Brunsden and John Thornes, and in the **geomorphic threshold** concept by Stanley Schumm. A landscape or geomorphic system is said to be sensitive if recovery from a major disturbing event, such as a major flood, is long-drawn-out in relation to the frequency of the disturbing event. In such landscapes there is a relatively

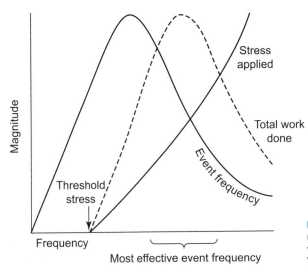

Figure 1.3 Relationships between event magnitude and frequency and geomorphic work done (modified from the classic work of Wolman and Miller).

high probability of further disturbance before the system has recovered from the previous event, for example if re-vegetation of eroded slopes is slow. On the other hand, landscapes that recover quickly from disturbance are said to be robust.

At a larger scale, a landscape may undergo sudden and rapid changes from one state to another, for example from stable to gullied hillslopes or from a meandering to a braided river channel. Such a landscape is said to have crossed a geomorphic threshold. These thresholds may be related to internal properties of the geomorphic system or may be brought about by external stresses; for example, by tectonic, climatic or human-induced changes in the environment. One research problem is to differentiate between externally and internally induced thresholds, a problem exacerbated by the common occurrence of a **complex response** to threshold-related changes to the geomorphic system. Robust landscapes are better able to withstand such stresses, but sensitive landscapes may be more susceptible to externally induced threshold changes. A modern consideration would be to identify landscapes that may be particularly vulnerable to change in response to global warming.

1.4 What we mean by the fundamental driving forces

The world's landforms are the result of the interaction between internal (geologically-driven) and external (climatically-driven) forces. The vast majority of the Earth's surface form is the result of climatically-driven erosional and depositional processes, operating on the surface of the planet, whose structure and composition are the result of geologically-driven internal processes.

1.4.1 Internal (geological) forces

Huge advances in our understanding of the Earth as a planet occurred in the late 1960s and 1970s through the development of the plate tectonics concept. In this model the Earth's crust is seen to be composed of a series of relatively stable rigid (lithospheric) plates separated from one another by less stable plate boundary zones.

The crust is lighter than the underlying mantle and rests on it in **isostatic** equilibrium; in other words, at an elevation proportional to the crustal thickness and density. The crust itself is of two sorts: lighter, thicker **continental crust**, composed dominantly of **granitic** rocks, and denser, thinner **oceanic crust**, composed dominantly of **basaltic** rocks, so the continents sit at higher elevations than the ocean floor. Most of the lithospheric plates include both oceanic and continental portions.

There are three types of **plate boundary**: constructive, destructive and conservative boundaries. Below the crust the upper mantle (the asthenosphere) is sufficiently near its pressure melting point to allow heat transfer by convection, and deformation by slow flowage. Convection within the upper mantle creates linear zones of upward heat transfer that partially melt the upper mantle (peridotite) rocks, releasing **basalt** lavas into the crust. These zones of upwelling are known as **constructive plate boundaries** as this is where new crust is formed by the extrusion of basalt lava. They form **mid-oceanic ridges**, separating two tectonic plates. The basalt lava is added to the inner margin of each plate, creating new oceanic crust, and thus widening the ocean basin in the process known as sea-floor spreading (Figure 1.4A). This process is slow; for example, the modern North Atlantic

A

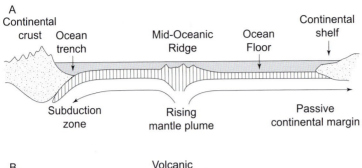

B

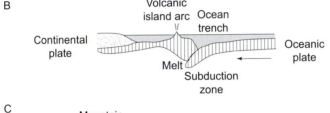

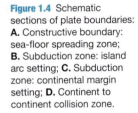

Figure 1.4 Schematic sections of plate boundaries: **A.** Constructive boundary: sea-floor spreading zone; **B.** Subduction zone: island arc setting; **C.** Subduction zone: continental margin setting; **D.** Continent to continent collision zone.

C

D

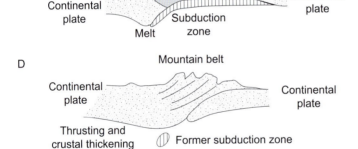

Ocean has developed over the last 70 million years or so. An early stage in this process is crustal rifting. The Red Sea rift system is beginning to separate the African and Arabian plates. Mid-oceanic ridges and rift systems are zones of volcanic activity.

At the same time as the ocean basins develop by sea-floor spreading, the opposite plate margin undergoes compression as the plate converges with a neighbouring plate This type of plate boundary is called a **destructive**

plate boundary, because it ultimately involves the destruction of crust and its absorption into the upper mantle. There are three types of convergent (destructive) plate boundary. In the first case the two plates are of oceanic crust, (Figure 1.4B) and the more mobile plate is forced below the other plate. In the second case, the opposing plate is formed of thicker continental crust (Figure 1.4C). The thinner, denser oceanic plate is forced under the continental plate. In both cases this process is called

subduction. The descending plate undergoes partial melting, causing volcanic activity at the surface. The third case occurs when two continental plates collide (Figure 1.4D). Both of the latter cases involve thicker continental crust and so considerable crustal thickening occurs, some due to subduction itself, but also as a result of the compressional tectonic setting forcing bodies of rock to be folded and to be thrust over one another.

Destructive plate boundaries have major topographic expression (Figure 1.5). Those involving oceanic crust are characterised by deep marine trenches and volcanic island arcs. Zones of crustal thickening are isostatically elevated to form mountain chains.

Continental-margin boundaries coincide with some of the Earth's major mountain chains, the 'young fold mountains', including those that more or less encircle the Pacific Ocean. Subduction below such mountain chains produces volcanic activity. Continent to continent collision zones result in the greatest amounts of crustal thickening, the greatest amounts of crustal isostatic uplift, and therefore form the highest mountains of all, the Himalayas.

Again, the timescales involved are enormous; for example, the evolution of the Western Cordillera of North America relates to westward movement of the Americas plate over the same timescale as the widening of the

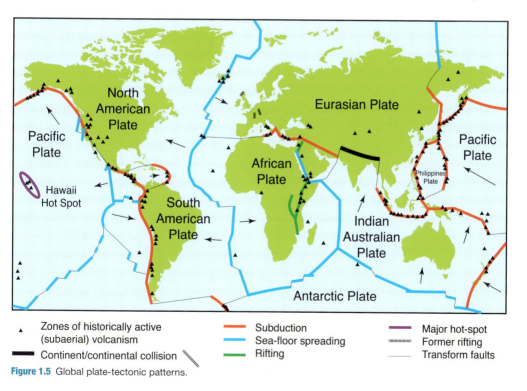

Figure 1.5 Global plate-tectonic patterns.

Atlantic Ocean (*see* above). Similarly the Eurasian Alpine/Himalayan mountain system has developed over a similar period as the result of the southern continental plates (Africa, India) encroaching on Eurasia and the closure of what was once an intervening Ocean (Tethys), roughly in the position of, but much larger than, the modern Mediterranean Sea.

A third type of plate boundary is a **conservative boundary**, where crust is neither created nor destroyed, but one plate moves laterally against the margin of another. These are major **transform fault** zones, and although earthquakes are common on all plate boundary types, transform faults are the sites of some of the most powerful earthquakes on Earth. Classic examples include the San Andreas Fault in California and the Anatolian Fault in Turkey.

The plate tectonics model provides a mechanism for the much earlier continental drift theory, within which, over geological time, the continents 'moved' in relation to one another. The plate tectonics model also provides the basis for interpreting previous geological patterns, as well as for understanding the modern patterns of gross spatial and elevational characteristics of the Earth's surface.

We will deal with the topographic expression of global-scale plate tectonics in more detail in Chapter 2, but there are also implications at the regional and local scales. At the regional scale (Chapter 3), the distribution of volcanic activity closely reflects the plate tectonics context together with the location of 'hot spots' above mantle plumes (Figure 1.5). The intensity of structural rock deformation by folds and faults also reflects modern and past plate tectonic activity. At the local scale the plate tectonics context is expressed by the location of individual volcanoes and local tectonic patterns.

1.4.2 External (climatic) forces

Gross relief (mountains, plains, etc.), can be related to plate tectonics, but the transformation of that gross relief into landforms is the result of processes largely generated by the climate system. Surface geomorphic processes (*see* Chapter 4), which can collectively be described as the **sediment cascade**, are driven largely by the climate system. These processes involve **weathering**, the breakdown of rock by mechanical and chemical processes, dependent on moisture and temperature; then erosion, transport and deposition of rock debris by various geomorphic systems, driven by gravity, flowing water including waves and currents, wind, and glacial ice. Apart from gravity, all of these are dependent on the climate system.

We will deal with the implications for global geomorphology in Chapter 2, but here we will focus on some of the climatic mechanisms important for geomorphic processes, particularly those related to temperature and moisture.

Temperature is important, both directly and in relation to moisture availability. A high diurnal temperature range affects heating and cooling of rock surfaces and therefore mechanical weathering. Similarly, the frequency of freeze-thaw activity affects the mechanical weathering regime. Sustained high temperatures and high moisture content accelerate chemical weathering processes. Mean annual temperature range is also important. Mean annual temperatures below $0\,^{\circ}C$ may result in a permanently frozen subsoil and bring about a whole range of soil and slope processes characteristic of **periglacial** environments. Sustained winter temperatures below 0 °C allow accumulation of an annual snowpack, which on melting in spring or summer may lead to heavy annual flooding. Sustained temperatures throughout the year below 0 °C allow perpetual snow accumulation and its conversion to glacial ice.

Moisture availability is also fundamental. Important is the relation between annual precipitation and annual potential **evapotranspiration**, which differentiates humid from sub-humid, semi-arid and arid environments. Where precipitation is in excess, soil moistures are maintained, groundwater recharge takes place and perennial rivers are sustained. In dry regions rivers are often ephemeral, but prone to flash floods from occasional storms. In humid regions on free-draining sites, high soil moistures promote rapid soil development, and the development of soil profiles is dominated by downward movement of moisture through the soil, whereas in dry regions soil formation is slower and there is much less downward movement. In humid areas, both temperate and tropical, soil formation tends to be faster than removal by erosion, leading to the general condition of soil-mantled landscapes. In arid areas the reverse is more common, with characteristically bare, eroded landscapes. Rainfall itself is an important geomorphic agent, especially high intensity rainfall, leading to erosion by run-off, to flood conditions, and often to the initiation of landslides.

Perhaps the most effective way of considering the impact of the climatic system on geomorphic processes is to consider the hydrological cycle. At a global scale the hydrological cycle involves transfer of water by evaporation, precipitation, glacial melt and river flows between the major water storage zones on the planet, the sea, the atmosphere, the land surface, underground, and glacial ice. During the Pleistocene ice ages, reduced global temperatures increased the proportion of the world's water stored in glacial ice, reducing that stored in the oceans, thereby reducing global sea levels (*see* Chapter 2).

However, it is at the regional and local scales, that of the drainage basin, that the hydrological cycle is most relevant for consideration of geomorphic processes. In Figure 1.6 the major stores of the drainage basin hydrological cycle are indicated by boxes, the flows by arrows, and the mechanisms controlling the flows by diamonds, with those processes that are of particular importance for geomorphology highlighted in bold.

Precipitation falls from atmospheric storage onto the land surface. If it falls as snow, it may be stored on the surface for some time as a snowpack, before it melts. If seasonal melting is rapid it can contribute significantly to rapid run-off and river flooding. If it falls as rain, the intensity of the rainfall is important in determining its subsequent behaviour. Vegetation will act as an umbrella, protecting the land surface below by **interception**, its effectiveness depending on the vegetation type and decreasing with increasing rain duration and intensity. The soil will absorb incoming rainfall at a particular rate, known as the **infiltration capacity**, depending on soil properties and on the antecedent soil moisture. In most cases infiltration capacity will be higher than the effective rain intensity, in which case almost all the incoming rain will be absorbed by the soil. Only when rain intensity exceeds the infiltration capacity will run-off (overland flow) take place.

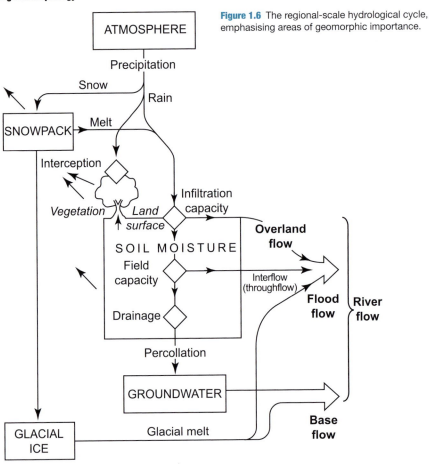

Figure 1.6 The regional-scale hydrological cycle, emphasising areas of geomorphic importance.

Some surfaces will have very low infiltration capacities (e.g. already saturated soils, frozen soils, very clayey soils, some bare rock surfaces, artificial concrete surfaces), in which case all incoming rain will run off. **Run-off** is an important geomorphic agent capable of hillslope erosion (*see* Chapter 4). It is also a major contributor to the **floodflow** component of the **hydrograph**, especially for flash floods.

This run-off and erosion model is most applicable in arid regions, where infiltration capacities are low and storm rain intensities tend to be high. In humid regions floodflows are more commonly fed by **saturation overland flow**, run-off from saturated soils. Such soils are usually located near the channel on low-angle, well vegetated slopes, and therefore yield much less sediment.

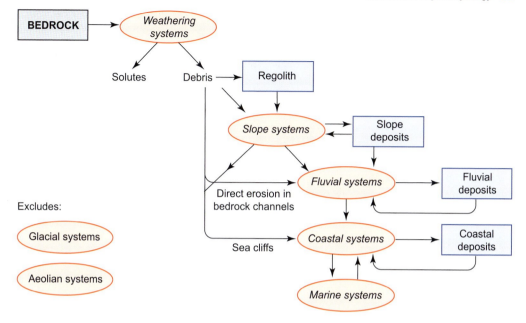

Figure 1.7 Schematic representation of the sediment cascade as applied to drainage basins (Note: for simplicity glacial and aeolian systems are excluded).

The fate of the soil moisture depends on another soil property, the **field capacity** of the soil, the amount of soil moisture that can be retained in the soil by **capillarity**, rather than draining downwards through the influence of gravity. If the infiltrating water does not bring the soil moisture up to field capacity, there will be no further movement; soil moisture will simply be used by plants or evaporate during dry weather. If, however, soil moisture exceeds field capacity, it will drain if it can, or if drainage is impeded, the increased water content will lead to a saturated soil. High soil moisture contents greatly reduce the stability of the soil, and may lead to shallow landsliding (*see* Chapter 4). Where drainage from the soil is possible, the excess water will drain

either laterally by what is known as **interflow**, or in free-draining soils vertically by percolation into the underlying bedrock to the water table as **groundwater**. Rapid interflow may be an important constituent of river flood-flow in humid areas. Groundwater will eventually drain through springs to river systems, sustaining the dry-weather flow (**baseflow**) of rivers.

The hydrological cycle operates very differently in different global climatic regions. In humid areas during rainy weather most of the cycle may operate. In arid areas, interception is limited and soils are shallow, limiting infiltration capacity; hence, during storm conditions run-off may be heavy. Also, soil moistures rarely reach field capacity, limiting groundwater recharge, as well as influencing pedological processes. There are therefore fundamental differences in geomorphic processes between humid and arid areas. Similarly, temperatures have a profound effect on the operation of the hydrological cycle, especially when freezing is involved. The seasonal presence of a snowpack or of frozen ground is of importance to geomorphology, as is the presence of long-term water storage as glacial ice. These processes again create major differences between global climatic regions, in terms of their geomorphic regimes.

1.5 Different approaches to the study of geomorphology

In studying geomorphology we are trying to do two things; first to explain how the landforms of the Earth developed (evolutionary geomorphology), and second, how landform processes operate (process geomorphology). In both cases it helps to consider geomorphic phenomena as systems that respond to energy and material inputs, in ways first suggested in the 1970s by Dick Chorley and Barbara Kennedy. Systems respond to variations in energy and material inputs by adjustments of their internal structure and morphology. In the first approach (evolutionary geomorphology) real timescales are an essential part of the system; the geomorphic system responds to real events (e.g. tectonic activity, glaciation, climatic change). Any concept of equilibrium is timebound, for example the initial high rates of erosion following tectonic uplift tend to lessen over time. In the second approach (process geomorphology) absolute time is less relevant. More important are aspects of time that relate to how the system responds to changes in inputs of energy or materials.

Under these conditions equilibrium concepts are less timebound. They relate to a balance between input and output and the internal adjustment of the system that maintains such a balance. For example, a river channel undergoes both erosion (sediment output) and deposition (sediment input). If the two are in balance then the channel morphology is likely to be maintained in a form of **dynamic equilibrium**. In other words, the channel morphology adjusts to variations of flood magnitude (energy input) and sediment supplied (material input).

There are two ways of considering such process systems; the first by considering the passage of energy and material through the system (as a cascade), the second by considering the internal structure of the system (as a morphological system). A cascading system comprises a series of flows between stores within the system. The hydrological system (*see* above, Section 1.4.2; Figure 1.6) is one such system relevant to geomorphology. The sediment cascade (*see* above, Section 1.4.2; Figure 1.7) is another. Mechanisms within the system control the flows from store to store. Sets of mechanisms can be viewed as cascading subsystems. Equilibrium conditions within any of the subsystems relate to a balance between inputs and outputs. An important aspect of understanding cascading systems is the awareness not only of the internal mechanisms, but also of the magnitude and frequency characteristics of the inputs to the system (*see* above, Section 1.3).

Morphological systems define the internal structure of a system and are usually described by relationships between the system components. These may be expressed statistically, but ideally relate to underlying causal relationships. For example, the **hydraulic geometry** of stream channels (*see* Section 4.3) expresses the relationships between flow and channel variables (e.g. by the response of channel width, depth, water velocity, and sediment transport properties to variations in water discharge). In many cases a network of causal relationships can be identified within morphological systems, which 'feed back' on one another. These can radically affect the equilibrium tendencies of the system. Some **feedback** relationships (known as negative feedback) tend to balance out the effects of disturbance. For example, in a river channel, bank erosion will tend to widen the channel, reducing depths, velocities and the erosional stresses on the banks, thereby reducing the likelihood of further erosion. On the other hand some feedback relationships (positive feedback) are self-reinforcing, and therefore tend to destabilise the system. For example,

mountain glacial erosion will deepen the **cirques** (*see* Section 4.5.2), which form the main gathering grounds for snow and ice, thus increasing the erosional potential. Ultimately only a major climatic change or complete erosional destruction of the lip of the cirque can break this vicious circle, in other words by the crossing of a major geomorphic threshold (*see* Section 1.3).

The two basic systems approaches (cascades and morphological systems) can be linked together in what is known as a process response system. Many of the cascade mechanisms can be treated as variables in morphological systems, allowing such linkages, and allowing the system as a whole to evolve through time. Such systems are of relevance not only for understanding process geomorphology, but also, because absolute timescales are involved, they can be applied to questions related to evolutionary geomorphology.

2 Global-scale geomorphology

In this chapter we consider global-scale geomorphology in relation both to the plate tectonics context and to global climatic patterns, present and past. Though dealing with global phenomena, we shall necessarily come down in scale when we consider examples of how these global phenomena are expressed in landforms. It is at this scale that the effects of plate tectonics on the form of the Earth's surface can be seen most clearly.

2.1 The plate tectonics context

2.1.1 The ocean floor

The form of the ocean floor is almost wholly determined by plate tectonics. Mid-oceanic ridges encircle the globe (Figure 1.5). They are formed by the injection of a magma chamber into the crust below zones of seafloor spreading. Away from the zone of seafloor spreading the oceanic crust forms extensive **abyssal plains**. Isolated seamounts and volcanic islands stand above the abyssal plains. These sit above stationary mantle plumes that form hot spots in the crust, triggering volcanic activity. The plumes themselves are geostationary, but the plates move over them, resulting in volcanic activity spanning a range of ages. For example, the Hawaiian islands sit over a hot spot in mid Pacific. The volcanoes in the north-western islands in the chain are now extinct, but Mauna Loa in the south-east is active, reflecting the north-westerly movement of the Pacific plate over the hot spot.

At the leading edge of a plate the ocean floor is modified by subduction. In oceanic subduction, a deep-ocean trench (Figures 2.1, 2.2A) is formed, beyond which lie island arcs, a situation found in many areas in south-east Asia and in the Caribbean. Continental margin subduction may also produce ocean trenches. On this type of continental margin, such as that on the west coast of the Americas, the coastline is relatively straight and there is no significant **continental shelf**.

At trailing-edge continental margins such as those around most of the Atlantic Ocean, the continental crust forms continental shelves, bounded by the **continental slope** down to the abyssal plain. At times of low global sea levels during Pleistocene glacial phases (Section 1.4.2), much of the continental shelf was exposed as dry land over which rivers flowed. Where the larger rivers (e.g. the Amazon, the Hudson) flowed over the continental slope some of them incised, cutting deep canyons. River-fed, sediment-charged water is funnelled down these submarine canyons as turbidity currents, spreading their sediment onto the sea floor as submarine fans. Near to land on the continental shelf, especially around the North Atlantic Ocean, there may be a relict terrestrial topography submerged by the post-glacial rise in sea level. This may be an erosional topography where the tops of the partially submerged ridges form islands (e.g. the Hebrides, off western Scotland), or a depositional topography, especially a glacial depositional topography, adding complexity to the morphology of the sea floor.

2.1.2 Global-scale continental landforms

At the global scale the topography of the continental areas also strongly reflects the influence of plate tectonics, but in a more complex way than that of the ocean floors. The young, high mountain ranges of the Alpine/Himalayan system and of the Pacific rim coincide with active or recently active destructive plate boundaries (compare Figures 1.5. and 2.1), their high elevation being the result of crustal thickening and sustained crustal isostatic uplift (*see* above, section 1.4.1). In many areas the detailed structures of these mountain ranges are far from simple. The broadly continent-to-continent collision that created the European Alpine system and the changing plate tectonics setting of the western United States have produced incredibly complex mountain systems at the regional scale (*see* Chapter 3).

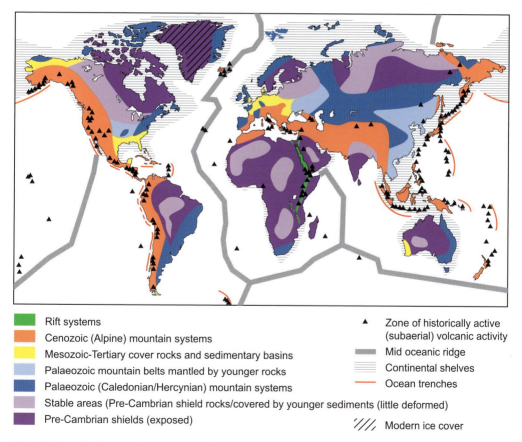

Figure 2.1 Global relief patterns and structural units.

Figure 2.2 Examples of satellite images of global/continental-scale features (©Google Earth).
A. The ocean floor off Indonesia: this is a subduction zone. Note the absence of a continental shelf, and the ocean trench off the west coast of Indonesia.

B. Deserts in central Australia. To the east are the folded rocks of the Flinders Ranges; in the centre are the salt flats of dry Lake Torrens; elsewhere are Pleistocene dune systems.

C. Parts of Arctic Canada and western Greenland. Note the extensive ice sheet cover, not only over Greenland to the east but also the smaller ice sheets on the Canadian Arctic islands.

In addition to young mountain systems created by the modern plate tectonics setting, the remnants of older mountain systems form lower, less dramatic mountains on every continent. These mountain systems relate to destructive plate boundaries that are no longer active. They owe their present relief in part to crustal isostatic uplift, related to the original crustal thickening, but their modern topography is purely erosional and related to the presence of older, harder, more erosionally resistant rocks. Two systems can be identified (Figure 2.1). The Hercynian/Variscan system dates from a major tectonic phase during Permian times, 300–250 million years ago (Figure 1.2). At that time the Atlantic Ocean did not exist and a major mountain system developed along the southern margin of a proto-Eurasian continent with an arm extending along the Urals. It extended westwards into what are now the Appalachians in the eastern part of the United States. In the southern hemisphere a similar system can be traced through eastern Australia, southern Africa and into southern Argentina. Today, only in a few places (e.g. the Appalachians, the Urals, eastern Australia) does the Hercynian system represent more or less continuous mountain chains. Elsewhere, especially in Europe, the Hercynian structures and their constituent rocks have been incorporated into younger Alpine structures, or the remnants of the Hercynian mountain system have been fragmented, by post-Hercynian faulting, into discrete upland blocks (e.g. the Massif Central in France), between which areas have subsided and been buried by younger sedimentary rocks.

The Caledonian mountain system relates to the closure of an ocean between the North American and Eurasian plates during the late Silurian and Devonian periods (c.400–350 million years ago, Figure 1.2). Today their remnant rocks and structures form the Scandinavian mountains, the upland areas of north-west Britain and much of Ireland. They can be traced across the Atlantic into North America, where they have been incorporated into the Appalachian structures. Another fragment is present in South America. Again, their elevation relates in part to the original crustal thickening, but their present relief forms are entirely erosional, related primarily to rock resistance.

Within each continent are zones that have been far away from plate boundaries and have been tectonically stable throughout the Phanerozoic. These are the cratonic '**shield**' areas (Figure 2.1), composed of Precambrian, mostly metamorphic, rocks older than 550 million years. Over extensive areas (e.g. the Baltic Shield, the Canadian Shield) the Precambrian rocks are exposed at the surface, but away from the core areas (under the Russian platform, under the North American plains) the Precambrian rocks are mantled by a little-deformed cover of younger sedimentary rocks. The shield areas of the northern continents (the Baltic Shield, the Canadian Shield) were heavily glaciated during the Pleistocene (*see* below, Section 2.2.2), therefore their detailed landforms are relatively young. However, the shield areas of the southern continents escaped glaciation during the Pleistocene, therefore they preserve ancient land surfaces dating back well into the Tertiary and possibly beyond.

The major lowlands of the continents are areas of less deformed younger sedimentary rocks that lie between shield areas, remnants of old mountain chains and modern mountain chains. In some cases these form distinct Mesozoic or Cenozoic sedimentary basins (e.g. the Paris Basin), but elsewhere they simply comprise relatively undeformed Mesozoic or Cenozoic sedimentary rocks burying older structures.

One other feature that can be related to modern and ancient plate tectonics patterns are rifts. These form over zones of mantle upwelling and may be the precursors of sea-floor spreading and the development of ocean basins. The East African rift system and its extension along the Dead Sea Rift constitute a modern zone of rifting, which eventually may mean the splitting of the African plate. In Europe an aborted discontinuous rift system, dating from the Miocene, can be identified in the Auvergne (France), and in the Rhine rift valley on the Franco-German border. The Eocene volcanic rocks of western Scotland and the Inner Hebrides may represent another early (Tertiary) aborted rift system that was replaced by the mid-Atlantic rift.

2.2 The global climatic context

2.2.1 Climatic geomorphology

The world's weather and climate patterns control the distribution of heat and moisture, which in turn control the distribution of geomorphic processes. We will deal with geomorphic processes in more detail in Chapter 4, but here we identify how different climatic regimes favour the operation of various geomorphic regimes. Box 2.1 summarises how different climatic regimes favour the operation of different geomorphic processes.

Climatic geomorphology, the basis of which is outlined in Box 2.1, deals with the climatically controlled distribution of geomorphic

Box 2.1 The geomorphic effectiveness of global climates

1. Arctic and Antarctic (Glacial) climates

In these areas mean annual temperatures are well below 0 °C.

Year by year snow accumulation converts to glacial ice.

Glacial processes are dominant.

2. Sub-arctic (Periglacial) climates

In these areas mean annual temperatures are below 0 °C.

The subsoil remains frozen (permafrost) throughout the year, but there is sufficient summer warmth to cause snowmelt floods, and to thaw the surface layers of the ground: slope processes (solifluction) are effective. Frequent freeze-thaw cycles and **nivation** are very effective in mechanical weathering.

3. Humid temperate climates

These climates cover a wide range of actual climates, but they have several characteristics in common. Precipitation exceeds potential evapotranspiration: soils are commonly moist, there is recharge of groundwater, sustaining perennial rivers.

Rainfall may be all-year (e.g. Western Europe), or seasonal with summer dominance (e.g. continental climates; American east-coast climates), or seasonal with winter dominance (the more humid Mediterranean climates). Summer temperatures may range from cool (e.g. Scotland) to hot (e.g. Italy, north-eastern USA) and winter temperatures may range from mild (e.g. Brittany) to cold (e.g. eastern Canada). Temperature and moisture conditions mean that the natural vegetation of this zone is forest, but much of this area is now used for agriculture. Under most 'natural' circumstances the rate of surface erosion is less than the rate of weathering and soil formation, so most landscapes are soil-mantled. The most effective geomorphic processes are slope (mass movement) and fluvial processes.

4. Dry climates

These climates again cover a wide range of conditions, but the critical aspect is that annual precipitation is markedly less than the high potential evapotranspiration. This results in dry soils, little groundwater recharge, and usually a scant vegetation cover. Summer temperatures are invariably hot; winters may range from mild to cold. Dry climates include semi-arid Mediterranean climates (e.g. Spain, Israel), interior continental steppe and dry grassland areas, desert margins and the truly arid climates of the great deserts of the world. Rainfall tends to occur in occasional heavy convectional storms, resulting in rapid run-off; therefore slopes are dominated by surface erosion rather than by mass movement processes. Rivers tend to be ephemeral, dry for much of the year, but responding rapidly to heavy rainfall by flash floods. Weathering is slow, but produces characteristic soils and surfaces. In truly arid areas, precipitation may be rare, and geomorphic processes may be dominated by wind action.

5. Tropical climates

In these areas temperatures are high all year round; freezing is virtually unknown, and precipitation increases towards the Equator from the great desert areas. Precipitation may be all-year as in the truly equatorial regions, but may be strongly seasonal, especially on the drier margins of the tropics (e.g. in the Sahel), and in Monsoon regions. In Monsoon regions wet-season rainfall may be exceptionally heavy. High temperatures and high precipitation favour rapid weathering and the development of very deep soil profiles. Natural vegetation ranges from grassland and dry woodland in the seasonally dry areas to tropical rainforest. In undisturbed areas geomorphic activity reflects the rainfall regime, with seasonally high river flows in the seasonally wet/dry regions to high perennial flows in the true humid tropics. Because of a naturally deep weathering mantle and the high rainfall, these regions may be prone to considerable human-induced disturbance.

6. Mountain climates

Because of their elevation, high mountain environments may experience very different climates from surrounding lowlands, generally with cooler temperatures and higher precipitation. Glacial conditions exist in high mountain areas in temperate and even in tropical latitudes. Mountain areas within deserts may well support forest vegetation. The geomorphology of mountain regions is made distinct not only by the climatic regime, but also by the presence of steep slopes, which tend to accelerate most geomorphic processes.

processes. Although modern processes essentially determine the gross landscape types (e.g. Deserts, the Arctic: Figure 2.2B, C,), in most regions the landforms carry a legacy of past processes, especially of those processes that were active during the Pleistocene.

2.2.2 Quaternary climatic change – glaciation

Over the last half million years or more the Earth's climates have oscillated between global glacial and interglacial conditions (*see* Section 1.3). In addition to the semi-permanent ice caps over Greenland and Antarctica, during the Pleistocene glaciations large continental ice sheets formed over large areas of the northern hemisphere, and glaciers in many mountain areas elsewhere became much more extensive. The limits of glaciation are fundamentally important for geomorphology in that they determine the spatial extent of past glacial and related processes. Two sets of limits are important; those related to the last glaciation (OIS2) reaching its maximum limits about 20,000 years ago, and those related to the maximum Pleistocene glacial extent that occurred during OIS6 or OIS8, >150,000 years ago.

Two huge continental-scale ice caps were formed in North America, the Laurentide and Cordilleran ice caps (Figure 2.3). At maximum glaciation the two ice sheets met in western Canada and their limits extended south to the latitudes of New York, the Ohio Valley, and in the west to somewhere south of the Canadian border. Separate mountain ice caps were present over some of the higher mountain ranges in the west, including parts of the American Rocky Mountains and the Sierra Nevada. Ice spread out from the centre of the Laurentide ice cap, scouring bedrock to produce the intensely eroded terrain on the Canadian Shield. Deposition, both by glacial ice and by temporary lakes during the melting of the ice, took place around the margins to produce the glacial depositional terrain of southern Canada and the Midwest of the USA. Similarly the Cordilleran ice cap and the smaller ice caps and glaciers in the other mountain areas tended to be erosional near-source in the mountains and depositional in the neighbouring lowlands.

In Europe there was a similar situation at maximum glaciation (Figure 2.3). The Scandinavian ice sheet was coalescent with the Scottish ice cap. Ice extended south to the Bristol Channel, London, across Holland into north Germany, Poland and around the Baltic. In and near the source areas, the Scottish Highlands, the Norwegian mountains and across the Baltic Shield, the ice was primarily erosional. Further from the source areas of the ice, across the English Midlands and the North European Plain, deposition was dominant. Further south, smaller ice caps and mountain glaciers occurred in the Alps and the Pyrenees.

In Asia there were ice caps over the Himalayas and in the mountains of north-east Asia. In the Southern Hemisphere, glaciation was much less extensive, but ice caps and mountain glaciers were present in the Andes, Tasmania and New Zealand.

Globally, the maximum glacial limits relate to a glaciation earlier than the last glaciation (Figure 2.3). This has important implications for the preservation of glacial landforms and their modification by non-glacial processes. This is nowhere better illustrated than in Britain (Figure 2.4, Box 2.2), where four zones can be identified.

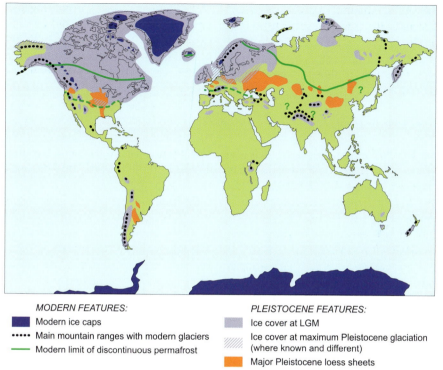

MODERN FEATURES:

- ■ Modern ice caps
- ••••• Main mountain ranges with modern glaciers
- — Modern limit of discontinuous permafrost

PLEISTOCENE FEATURES:

- ▨ Ice cover at LGM
- ▧ Ice cover at maximum Pleistocene glaciation (where known and different)
- ■ Major Pleistocene loess sheets
- - - - Southern limit of features related to Pleistocene permafrost (where known)
- ? Extent of Pleistocene permafrost uncertain

Figure 2.3 Global extent of modern and Pleistocene cold-climate phenomena.

Box 2.2 Zonation of British landscapes in relation to the glacial limits

The first zone is southern England. With the exception of a small area on the north Devon coast, and possibly of the Scilly Isles, the area south of the Bristol Channel and of the Thames valley has never been glaciated. In this area there is a greater preservation of landscape patterns inherited from the late Tertiary (*see* Section 1.3). River system development continued throughout the Pleistocene, without interruption from glaciation. However, this area experienced **periglacial** processes (hillslope processes influenced by the presence of permafrost; *see* Section 4.2) through all the glacial phases of the Pleistocene. The hillslopes have been extensively smoothed by **solifluction** (Figure 2.5A), and great thicknesses of **head deposits** (*see* Figure 4.7C) have accumulated, particularly on the valley sides.

The second zone is that between the maximum glacial limit and that of the last glaciation, essentially midland and eastern England. This area was glaciated, but more than 150,000 years ago; hence

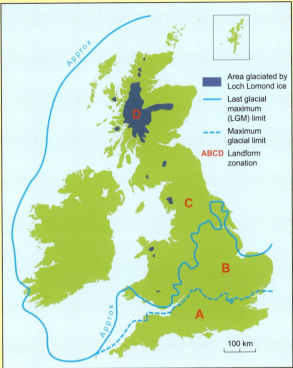

Figure 2.4 Zonation of British landscapes in relation to Pleistocene glacial limits. Zone A has never experienced glaciation. Zone B was glaciated >150 ka ago, but was ice free throughout the last glaciation, so was affected by periglacial processes throughout the last glacial period. Zone C was glaciated at the Last Glacial Maximum (LGM), at 20 ka, but the ice had melted completely by 15–13 ka. The glacial forms were then affected by periglacial processes during the late Pleistocene. Zone D experienced glacial ice cover for about 500 years during the Loch Lomond (Younger Dryas) glaciation, the ice melting rapidly as climate warmed into the temperate Holocene 10 ka ago.

Legend (on map):
- Area glaciated by Loch Lomond ice
- Last glacial maximum (LGM) limit
- Maximum glacial limit
- **ABCD** Landform zonation

100 km

the drainage pattern has been deranged by glaciation. The predominantly glacial depositional topography was modified by periglacial processes throughout the last glaciation and preserves little of the original depositional form.

The third zone is most of the area within the last glacial limit, including most of Wales, everywhere north of a line from Shropshire to the Humber area being glacial, with the exception of the Peak District of Derbyshire and the North York Moors. Glaciers reached this limit at the Last Glacial Maximum (LGM) about 18,000 years ago. The ice melted last of all in the Spey valley of Scotland about 13,000 years ago. In most of this zone

glacial erosional landforms (*see* Chapter 4) characterise the highest areas, with glacial depositional landforms elsewhere. The glacial landforms are relatively fresh, though they have been modified by varying durations of periglacial processes that occurred during cold conditions after the melting of the ice cap but before the end of the Pleistocene 10,000 years ago (Figure 2.5B).

There is one other limit that we need to take into account, which defines the fourth zone. This is the limit of a small ice cap that developed at the very end of the last glaciation around 10,000 years ago, and persisted for around 500 years, the so-called Loch Lomond Readvance (the

Figure 2.5 British landscapes (representative of the zones defined in Figure 2.4). **A.** *Zone A:* a periglacial landscape in an area never glaciated, the Quantock Hills, Somerset. Note the strongly convex hillslopes, characteristic of periglacial solifluction processes (*see* Section 4.2). Note also the Holocene incision into the valley floor. **B.** *Zone C:* a landscape glaciated during the LGM, but ice free since 13 ka. Yarrow Valley, Southern Uplands of Scotland. This landscape bears the imprints of both glaciation (deep scour of the valley into the upland plateau; deposition of glacial till – *see* Section 4.5) and limited periglacial processes during the last several thousand years of the Pleistocene. Note the extensive solifluction surface in the centre of the photo, trimmed at the base by Holocene fluvial activity. **C.** *Zone D:* A landscape glaciated during the Loch Lomond stage 10 ka, with little **periglacial** modification, but modification under **paraglacial** conditions by slope and fluvial processes during the Holocene. Glencoe, Western Highlands of Scotland.

Younger Dryas period, in European terminology). It occupied the southern and western Grampian Highlands, with other glaciers in the Cairngorm Mountains and the north-west Highlands, plus small glaciers in the Southern Uplands, the English Lake District and Snowdonia in north Wales. The Loch Lomond phase ended with a rapid climatic warming, so there was no periglacial modification of the glacial topography. Within the Loch Lomond limits the glacial erosional and depositional topography is fresh (Figure 2.5C).

2.2.3 Other changes to the climatic system during the Pleistocene

Outside the global glacial limits there were other climatic changes related to changes in atmospheric circulation. During glacial periods there had been a marked reduction of global temperatures, allowing the development of permafrost and the operation of periglacial processes southwards into the USA and into Europe (Figure 2.3). Even where there was no permafrost, in such areas as the American West, and Spain, freeze-thaw activity was much more effective than it is today. Another effect, in the periglacial regions around the margins of the continental ice sheets, was deposition of windblown silt (**loess**). Loess blankets the topography in some parts of the American Midwest, and across northern Europe in a belt through Belgium and Germany, but the greatest thickness of loess deposits is in the Loess Plateau in central China (Figure 2.3). Interbedded with the Chinese loess deposits are a series of interglacial **palaeosols**, the whole sequence preserving a complete palaeoclimatic record for the whole of the Pleistocene.

With the reduction in global temperature there was also a reduction in evaporation and a consequent reduction in precipitation. This produced a cold, dry steppe climate in the Western Mediterranean region, and greater aridity in some of the world's dry regions, probably so in parts of Australia and southern Africa. Elsewhere some of today's deserts were cooler and more humid, allowing rivers and lakes to exist in, for example, the American south-west and the Sahara. Indeed, desiccation of the Sahara occurred only in the mid-Holocene, well into the present interglacial.

2.2.4 Quaternary sea-level change

During global glacials a much higher proportion of the world's water was stored within the continental ice sheets rather than in the oceans, allowing world-wide sea levels to fall by about 100 m (**eustatic sea-level change**). The continental shelves were exposed; many islands (including Britain) became continental peninsulas; 'land bridges' were created (e.g. between Alaska and Siberia). Around the great ice sheets, the picture was complicated by the crust being depressed under the weight of ice (**isostatic sea-level change**).

On deglaciation, global eustatic sea level rose from its lowest point about 18,000 years before present (BP), coincident with timing of the last glacial maximum, until about 6000 BP when the last ice melted from the Laurentian ice sheet (Figure 2.6). Hence modern coasts are young features – less than 6000 years old. Adjacent to the glaciated areas the picture was complicated by the isostatic rebound of the crust as the weight of the ice sheets was removed (Figure 2.7). In such areas eustatic

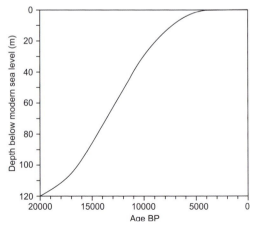

Figure 2.6 Post-glacial eustatic sea-level curve.

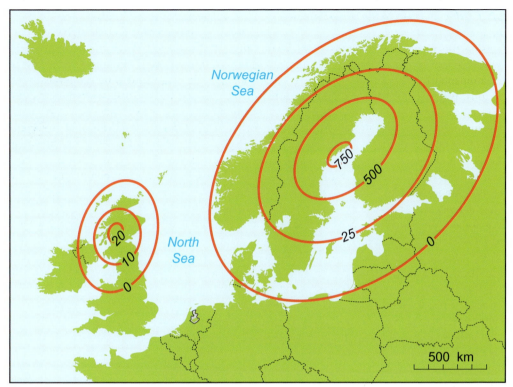

Figure 2.7 Europe: extent of glacio-isostatic rebound (contours in metres) since the melting of the ice sheets from the Last Glacial Maximum.

and isostatic processes interacted. Eustatic inundation was an immediate response to the water released from the melting ice sheets, but isostatic rebound is still going on today. Late-glacial beaches were formed on the west coast of Scotland (Figure 2.8A) as sea level rose and inundated the still depressed coastline. They became raised beaches as the rate of isostatic rebound overtook the rate of sea-level rise. A most impressive sequence of such raised beaches is around Hudson's Bay, depressed under the centre of the Laurentide ice sheet (Figure 2.8B). There is a complex sequence around the Baltic Sea and a similar one in

eastern Canada. There, as the Laurentide ice sheet melted, the sea flooded the St Laurence Lowland at a time when eustatic sea levels were still about 50 m below modern sea levels, but the crust was depressed below that level. Since then, isostatic rebound has elevated the saline clay sediments deposited by the 'Champlain Sea' to about 200 m above modern sea levels.

In summary: over the last 500,000 years or so, glacials have tended to last longer than interglacials. We live during an interglacial (the Holocene) which has lasted roughly for the last 10,000 years – and in the absence of

Figure 2.8 Glacio-isostatic modification of shorelines. **A.** Late Pleistocene raised coastal platforms, Loch Linnhe, western Scotland. The flat surfaces of the two islands in the centre of the photo represent two late Quaternary raised coastal platforms. **B.** Satellite image (from ©Google Earth) of the shore of Hudson's Bay, Canada. Note the sub-parallel beach ridges inland from the present shoreline.

any human-induced global warming, could be expected to last for maybe another 10,000 years or so.

What is important, from the point of view of geomorphology, is that present conditions have persisted for approximately only 10,000 years. Most of our landscapes bear the imprint of past conditions, either directly of glaciation or of the effects of previous climates. This has a two-fold importance for geomorphology. Many landforms are relict features, only partially adjusted to present day conditions. However, the landforms and their constituent sediments may preserve evidence for past climatic regimes.

2.3 Global-scale interactions between tectonic and climatic forces

In this chapter so far, we have seen how global plate tectonic patterns and global climatic patterns, past and present, have influenced global geomorphology. One way in which the interaction between global tectonic and climatic forces is expressed is in the relation between uplift and denudation rates. In zones of modern or past destructive plate margins considerable crustal thickening occurs. The relatively light continental crust rises isostatically, giving mountain ranges their high elevation (*see* Section 1.4.1). The high elevations create steep gradients, stimulating rapid incision by the drainage network. This in turn stimulates high erosion rates, which reduce the mass of the mountain range, stimulating further isostatic uplift. Recent studies of crustal uplift rates, involving complex geophysical methods, have demonstrated a general relationship between uplift rates and estimates of denudation rates, derived from river sediment load data (*see* below). Uplift continues, albeit at a diminishing rate, long after plate-tectonic activity has ceased, because the crustal thickening persists. This 'post-orogenic' process is known as **epeirogenic** uplift and affects large regions, particularly former mountain systems. There are other possible mechanisms, related to mantle processes, that cause epeirogenic uplift. For example, it is thought that some of the relief patterns on the African continent, a long way from past or present plate boundaries, are related to mantle processes. Similarly the sustained uplift of the Colorado Plateau in the American south-west is thought to be due to mantle processes. Indeed, epeirogenic uplift, driven isostatically by erosional offloading is now seen as a possible mechanisms for the continued uplift of continental areas, not only in mountain areas with significant crustal thickening, but also to a

Table 2.1

The "top ten" of the worlds river systems, ranked a) in terms of drainage area, b) river discharge, c) total river sediment load and d) specific sediment load (load /drainage area), the latter set of rivers taken only from large rivers with drainage areas > 500,000 km2 (data source – D Higgitt).

BY DRAINAGE AREA (10^6 km²)	BY DISCHARGE (km³yr⁻¹)	BY TOTAL SEDIMENT LOAD (10^6t yr⁻¹)	BY SPECIFIC SEDIMENT LOAD (t km⁻² yr⁻¹)
1. Amazon 6.15	Amazon 6307	Amazon, 1150	Yellow 1102
2. Congo 3.70	Congo 1290	Yellow 1080	Irrawaddy 888
3. Mississippi 3.34	Parana 1101	Ganges 524	Brahamaputra 852
4. Nile 2.72	Orinoco 1101	Brahamaputra 520	Magdelena 846
5. Parana 2.60	Yangtze 899	Yangtze 480	Ganges 535
6. Yenisei 2.58	Mississippi 580	Mississippi 400	Indus 260
7. Ob 2.50	Yenisei 561	Irrawaddy 364	Yangtze 247
8. Lena 2.43	Lena 511	Indus 250	Mekong 198
9. Yangtze 1.94	Mekong 470	Magdelena 220	Amazon 187
10. Amur 1.86	St. Lawrence 451	Mekong 160	Pearl 174

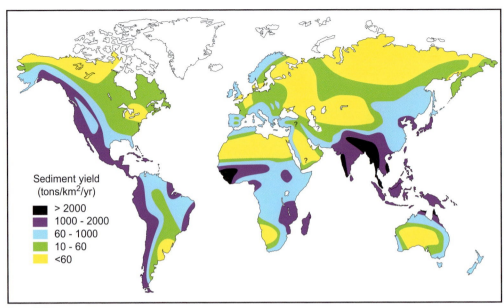

Figure 2.9 World sediment yields.

lesser extent away from mountain areas. The uplift is compensated for by flow of material in the lower crust below the uplifted areas. The reverse appears to take place under crustal loading in subsiding sedimentary basins.

Another way in which the interaction between global tectonic and climatic forces is expressed is in the global patterns of river sediment yield from the continents to the oceans. River sediment loads are a rather crude measure of net erosional amounts within their drainage basins, disregarding within-basin sediment storage. However, they do throw some light on the interactions between the two sets of driving forces. Obviously, in general the largest rivers tend to deliver the largest amounts of sediment to the oceans (Table 2.1). There are, however, some interesting anomalies that reflect the interplay of tectonic and climatic forces. Very high sediment yields are indicative of high rates of erosion. For example, the Yellow River drains the highly erosional and deeply dissected loess plateau of central China, a reflection of the interplay between uplift and Pleistocene climatic conditions. When sediment yields are considered in relation to drainage basin areas (Table 2.1; Figure 2.9), the rivers of southern Asia stand out. These rivers drain the Himalayan mountain belt within a broadly monsoonal climate. High erosion rates here are produced by a combination of rapid tectonic uplift with its associated deep dissection, and climatically-led factors producing intense seasonal precipitation.

3 Regional-scale geomorphology

If we come down in scale from the continental to the regional, global controls still exist in the background, but their status changes to one of context, within which local structure, local patterns of resistance to erosion, and local responses to climatic change become more significant.

It is at this scale that most people become aware of geomorphology; the scale that ranges from the sub-continental (e.g. the Alps, the Scottish Highlands, the Colorado Plateau) to the 'landscape' scale of hills and valley systems. This scale has received less attention in modern research than either the global/continental or the local scales. However, this is the scale that dominated traditional geomorphological research during the first half of the twentieth century. Much of that research was based on the ideas of a pioneering American geomorphologist, W. M. Davis, working at the turn of the nineteenth and twentieth centuries. Many of his ideas have since been demonstrated to be hopelessly simplistic, but some are still valid today. Davis realised that landscapes were the product of the interaction of three sets of factors: structure, process and time. We will use these headings, but in a modern context, as the framework for this chapter.

3.1 Regional scale – structure

By structure we mean two things: (i) the overall geological (plate-tectonic) setting and its influence on the landscape, and (ii) more specifically the influence of the underlying materials on surface form.

3.1.1 The regional-scale plate-tectonic setting

The plate-tectonic setting is very obvious when we consider landscapes at the regional scale, and is expressed in the differences between, for example, mountain ranges, plateaux and lowlands. Of fundamental importance is the overall relief, related to either active tectonics, or epeirogenic uplift (*see* Section 2.3). The difference between the overall relief and the regional **base level** governs the amount and rate of incision of the drainage network and hence slope steepness. Base level is the level below which incision cannot occur, which in many cases would be sea level, though more locally it may be the elevation of a major valley floor or of a zone of resistant rock. Ongoing or pulsed uplift may trigger renewed incision, resulting in '**rejuvenation**' of the landscape. This might be expressed by steep inner valley slopes, incised gorges, knick points and other irregularities in stream long profiles (*see* below, Section 3.2.1).

Other expressions of the regional geological setting include the direct creation of relief features by internal processes (*see* Section 3.1.2 below), the structural disposition of different rock types affecting resistance to erosion (Section 3.1.3), and the topographic expression of structure (Section 3.1.4).

3.1.2 The direct creation of relief by internal processes

Ongoing tectonic and volcanic activity can create landforms directly (e.g. fault scarps, volcanoes; *see* Figure 3.1), but these are rapidly modified by erosional processes, and are restricted in occurrence to areas of ongoing tectonics and volcanic activity respectively (*see* Section 1.4.1).

3.1.3 Lithology – resistance to erosion

Different rock types have differing degrees of resistance to erosion. Of the three rock groups, (**igneous, sedimentary and metamorphic rocks**), igneous rocks are mostly crystalline and hard, therefore resistant to mechanical breakdown and erosion. Lacking the bedding characteristic of sedimentary rocks, their only major internal weaknesses are joints and fractures, along which weathering proceeds (Figure 3.2). Their major weakness is that many of their constituent minerals may be susceptible to chemical change (*see* Section 4.1), so that in areas of intense chemical weathering (e.g. the humid tropics), these rocks may break down rapidly. Otherwise they tend to form areas of higher relief than the surrounding (weaker) rocks.

Figure 3.1 Direct relief creation by volcanic and tectonic processes. **A.** Fresh spatter cone and lava field, erupted less than 250 years ago, Lanzerote, Canary Islands. **B.** Active faulted mountain front, Panamint Valley, California, USA. Note the triangular slope facets on the spur ends indicating recent faulting, and the complex of alluvial fans issuing from the mountain catchments (*see* Section 4.3.4.2).

Figure 3.2 Influence of lithology on relief. **A.** Granite terrain, Central Mountain system, Spain. Note the joint patterns in the fractured rock and the rough surface topography. **B.** Badland terrain, cut in weak marls, northern Provence, France. Note the resistant **caprock** and the deeply dissected gully systems on the hillslopes.

Sedimentary rocks include a great range from weak clays and marls (Figure 3.2B) to mechanically stronger sandstones and limestones. They are bedded so that the bedding planes provide zones of weakness, facilitating weathering and erosion. Quite often relatively weak and strong rocks are interbedded, so that the weaker layers are eroded first.

Limestones are a particular case that have generated a whole sub-field of study, '**karst** geomorphology', named after the classic 'Karst' area in the Dinaric Alps of Slovenia.

Although most limestones are often mechanically strong, they are composed of calcium carbonate ($CaCO_3$, **calcite**), which is soluble in weak acids (rainwater, and humic acids derived from the soil), so they are susceptible to **solution**. Limestones are distinct in that the geomorphic processes are dominated by solution. The surface of limestone outcrops often exhibits small-scale solutional features, such as grooves and ridges, known as 'clints and grykes'. Where solution has taken place beneath a mat of soil or vegetation,

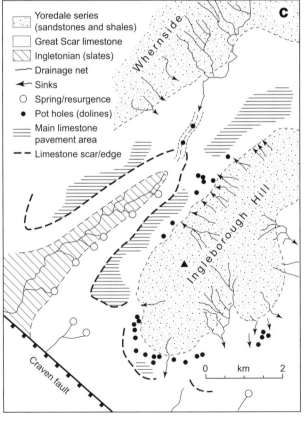

Figure 3.3 Limestone terrain. **A.** Limestone pavement, Malham, Yorkshire, England. Note the angular joints, but the rounded rill margins on this pavement formed under a soil cover. **B.** 'Rinnenkarren', Majorca. Note the sharp rill edges formed by surface run-off. **C.** Map of the Ingleborough karst, Yorkshire. Note how the streams that rise on the sandstone and shale geology of the slopes of Ingleborough Hill sink underground through swallow holes at the margins of the limestone outcrop, and reappear as springs in the valley bottom at the base of the limestone.

Legend (Figure 3.3C):
- Yoredale series (sandstones and shales)
- Great Scar limestone
- Ingletonian (slates)
- Drainage net
- Sinks
- O Spring/resurgence
- • Pot holes (dolines)
- Main limestone pavement area
- Limestone scar/edge

Whernside

Ingleborough Hill

Craven fault

0 km 2

as is common on the **limestone pavements** of Britain and other humid areas, especially those glaciated during the Pleistocene (Figure 3.3A), the rock edges are usually smooth. Alternatively sharp edged 'rinnenkarren' features are produced by solution from run-off over the surface (Figure 3.3B). Below the surface, joints within the rock are enlarged, allowing surface water to pass underground. Rivers and streams entering limestone terrain often pass underground through 'swallow holes'. A characteristic of most limestone areas is the absence of surface streams (Figure 3.3C). Underground, cave systems develop as water moves laterally along bedding planes and vertically along joints down to the water table. River resurgences (**Vauclusian springs**) occur at the base of the limestone or where the water table intersects the surface. Cave systems may eventually collapse to form small surface depressions (**dolines**) or much larger features (**polje** and **uvula**).

It is difficult to generalise about metamorphic rocks (those formed from the original sedimentary or igneous rocks by intense heat and/or pressure), because their character varies with the grade of metamorphism. Higher grade metamorphic rocks are crystalline, and therefore may be mechanically strong, but they include minerals that may be susceptible to chemical weathering. Their resistance to weathering and erosion may resemble that of igneous rocks, but metamorphic rocks may have a slaty cleavage, be foliated or banded, all properties that may accelerate their decay.

During the erosional development of the landscape, the weaker rocks tend to be eroded more rapidly, leaving the terrain on the resistant rocks upstanding to form the hill areas. In areas of contrasting rock resistance, the spatial patterns of relief therefore reflect the underlying geological structure (*see* below, Section 3.1.3). In areas of massive rocks the landforms may pick out not so much the contrast between neighbouring rock bodies, but the weaknesses within the rocks, such as faults and joint and fracture patterns.

3.1.4 *Topographic expression of geological structures*

Geological structures are the result of past tectonic activity, and include relatively simple fault and fold structures, complex structures related to igneous rocks, and the even more complex structures of mountain chains. The topographic expression of geological structure is not directly the result of the structures themselves, but of how they affect the disposition of rocks, or zones within rocks of differential erosional resistance. The fact that the rocks forming Snowdon Mountain in north Wales are a folded sequence of Ordovician volcanic rocks is almost irrelevant to the geomorphology of the mountain, which owes its form to glacial erosion of these volcanic rocks during the Pleistocene.

Faults are fracture planes within rock, along which the two bodies of rock on either side of the fault plane have moved against one another. The movement may result from tensional stresses (a normal fault, in which case one body of rock moves down along a sloping fault plane), compressional stresses (a reverse fault, in which case one body of rock moves up a low to steep angle fault plane), or lateral stresses (a strike-slip fault, in which case the movement is horizontal along a vertical or sub-vertical fault plane). The fault may have a direct topographic expression (*see* Section 3.1.1), either to form a fault scarp, or in the case of strike-slip faults, to laterally offset topographic features, such as stream channels. These features are present only in tectonically active areas. A well-known example is the offset drainage along the San Andreas fault in California. More commonly the geomorphic role of faults is either as a relatively easily eroded zone of shattered rock, or to bring two rock types of differing erosional resistance adjacent to one another. On erosion of the weaker rock, a fault-line scarp may be produced.

Terrain on simple **uniclinally** dipping sedimentary rocks may lead to scarplands, where the more resistant rocks form escarpments,

asymmetric ridges comprising scarp and dip slopes (Figure 3.4). The detailed form of escarpments will depend on the thickness of the resistant bands of rock, the dip of the strata and the overall available relief. There are numerous examples in southern England (North and South Downs, the Chilterns, the Cotswolds) and northern France. More complex is terrain on folded sedimentary rocks, involving outfacing and infacing escarpments on either side of **synclinal** or breached **anticlinal** structures respectively. In this type

Figure 3.4 Escarpment morphology. **A.** Diagram to illustrate escarpment geology and morphology. **B.** Escarpment of the Swabian Alb, southern Germany, an escarpment formed by Jurassic limestones, dip of the strata is to the east (left).

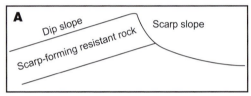

A

Dip slope

Scarp slope

Scarp-forming resistant rock

B

of terrain both **resequent** (anticlinal ridges, synclinal valleys) and **inverted relief** (anticlinal valleys, synclinal ridges) are possible (*see* Figure 3.5)

In more complex structures, again the form of the terrain developed generally has little to do with the creation of the structures themselves, but simply reflects the differing erosional resistance of the rocks involved. In major mountain chains the rocks may be deformed by complex overfolds, known as **nappes**, but the form of the relief owes much less to the structural form than to uplift, dissection, and rock resistance to erosion.

The same is true of igneous intrusions. For example, in Dartmoor in southwest England, the relief patterns simply reflect the differential erosional resistance of the Dartmoor granite, the surrounding aureole of metamorphic rocks and the shaly rocks of the country around. To a large extent this is also true of the complex structures of mountain belts, although the overall relief does reflect the

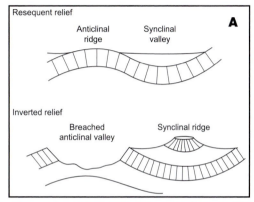

Figure 3.5 Folded terrain. **A.** Diagram illustrating relations between structure and topography in areas of resequent and inverted relief. **B.** Air view of a breached anticline in the Zagros Mountains, Iran. Note the transverse drainage- this has been interpreted as either antecedent drainage or superimposed from shale horizons above the scarp-forming resistant rocks (see Section 3.2.1).

amount of tectonic uplift, or in the case of ancient mountain systems, ongoing crustal isostatic uplift.

3.2 Regional scale processes – the drainage network.

We shall deal with the details of geomorphic process/landform relationships in Chapter 4, but here we need to consider the overall erosional development of the landscape, from tectonic uplift to the development of a drainage network and its incision into the underlying rocks.

3.2.1 Drainage evolution

Regional drainage patterns and their relationships to the underlying structure reveal a great deal about the long-term history of landscape evolution. Drainage is initiated in relation to the gradients produced by the original patterns of uplift. This drainage (**consequent drainage**) may exhibit a parallel or radial pattern, with local convergence into a branching network to form a **dendritic** drainage pattern (Figure 3.6). Provided the gradient is sufficient, the streams incise into the underlying bedrock. The rate of incision is controlled by: (i) the regional base level (*see* Section 3.1.1, above); (ii) stream power (related to the product of gradient and flood flow volume) – this, together with the distal base-level control, tends to produce a concave upward longitudinal stream profile – and (iii) rock resistance to erosion (which may modify the stream profile).

As incision takes place, the hillslopes, especially those facing towards the main stream, are modified and steepened, promoting new lines of convergent drainage (**subsequent drainage**). In simple uniclinally dipping sedimentary terrain, where these pick out bands of weaker rock running more or less orthogonal to the main stream, these may be preferentially eroded, producing a **trellised** drainage pattern (Figure 3.6).

Where the subsequent streams cut back and intercept earlier lines of consequent drainage, they may capture that drainage. **River capture** has important implications. Downstream of the point of capture the captor stream will increase its discharge and therefore enhance its stream power; the beheaded stream will lose power. At the capture point the bed of the original consequent stream will be lowered, creating a new lower local base level, which may cause a wave of incision to work headwards through the system.

In this way adjustment of the drainage pattern to structure takes place by river capture (Figure 3.6), gradually replacing the original consequent drainage pattern by a subsequent pattern. A small scale incipient capture is illustrated on Figure 3.7. A small consequent drainage flowing down the dip of resistant strata is about to be captured by a deeply incised subsequent stream that is aligned along the strike of underlying highly erodible marl. A major capture in the same area of south-east Spain is considered in more detail in Chapter 5.

There are, however, complications to simple sequences. Further uplift, tectonic deformation or a fall in regional base level may interrupt the development, triggering a new wave of incision, **rejuvenating** the system. This is often expressed by an incised drainage network set below a more gently sloping former landscape.

There are several other ways in which a lack of adjustment between drainage pattern and structure may occur (Figure 3.8):

Consequent Patterns

Dendritic pattern   Parallel pattern   Radial pattern   Centripetal pattern

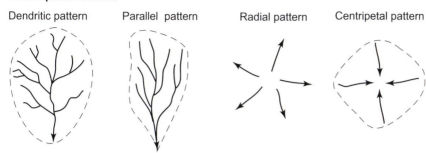

Sunsequent Patterns

Trellis pattern

Capture site

Beheaded consequent

Original consequent stream

Subsequent stream

Deranged drainage
(e.g. after glaciation)

Figure 3.6 Drainage patterns. **Above:** consequent patterns: dendritic, parallel, radial, convergent patterns. **Below left:** subsequent patterns: trellis pattern. Note the river capture. **Below right:** deranged irregular drainage.

Figure 3.7 River capture: an example of incipient stream capture from south-east Spain on the flanks of Catona Hill, Almeria. The valley running obliquely down the right-hand side of the photo (the valley floor has been modified by agricultural terracing) is the original northward-flowing drainage. This drainage is about to be captured by the drainage to the east (left) of the photo, marked by the steep gullying at the head of the drainage.

A   Superimposed drainage

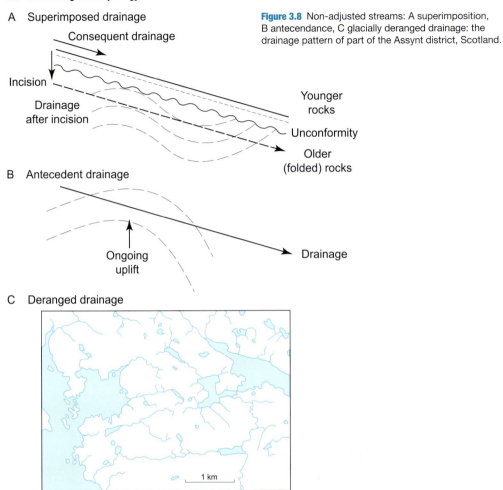

Figure 3.8 Non-adjusted streams: A superimposition,
B antecendance, C glacially deranged drainage: the
drainage pattern of part of the Assynt district, Scotland.

1) Consequent streams initiated on a cover rock may cut down through a geological **unconformity** into underlying rocks with a very different structure – this results in a **superimposed** drainage pattern in which the drainage is transverse to the major structures.

2) Another cause of transverse drainage may occur in tectonically active regions where uplift takes place across the line of drainage, and providing that the river maintains a rate of incision faster than the rate of uplift, may result in the river crossing a fold axis. This results in **antecedent** drainage.

3) Glaciation may radically disrupt the drainage pattern (**deranged drainage**), causing river diversions, or on glacial depositional surfaces, a totally new (often random) drainage pattern may develop.

3.2.2 *Drainage network composition*

There is a completely different approach to the study of drainage networks: a functional approach, as opposed to the evolutionary approach outlined above. This approach, first developed by Robert Horton in the 1940s, then elaborated by Arthur Strahler in the 1950s, depends on the hierarchical classification of the components of the drainage net. Under the Strahler system unbranched headwater streams are defined as first-order streams, two first-order streams join to form a second-order stream, and so on (Figure 3.9). A mature network accords with the first two laws of drainage composition; first, that the number of streams shows an inverse geometric

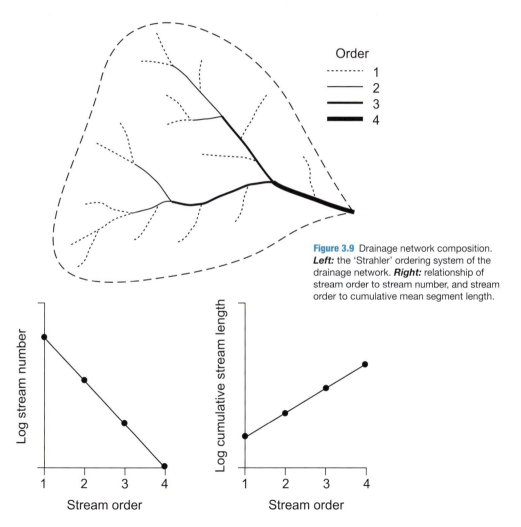

Figure 3.9 Drainage network composition. *Left:* the 'Strahler' ordering system of the drainage network. *Right:* relationship of stream order to stream number, and stream order to cumulative mean segment length.

relationship in relation to **stream order**; and second, that cumulative stream lengths show a direct geometric relationship with stream order (Figure 3.9). Immature or disrupted networks will depart from these idealised relationships.

The classification of stream segments in this way provides a basis for the quantitative study of the **morphometry** of streams and their drainage basins, considering not only stream numbers and lengths, but also gradients and drainage basin characteristics such as area, shape and slope properties.

3.3 Regional scale – evolution

Implicit in the evolutionary study of landforms is the polygenetic origin of landscapes. Landscapes change and develop through time, and preserve evidence of previous conditions. The first real attempt to systematise this concept was the so-called 'cycle of erosion' developed by W. M. Davis (Figure 3.10) in which, following initial uplift, landscapes went through a series of stages, termed 'youth, maturity and old age'. A youthful landscape would preserve much of the original (pre-uplift) surface, into which steep streams are deeply incised. By

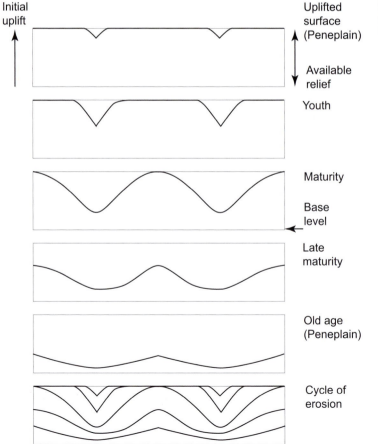

Initial uplift

Uplifted surface (Peneplain)

Available relief

Youth

Maturity

Base level

Late maturity

Old age (Peneplain)

Cycle of erosion

Figure 3.10 Schematic representation of the 'Davisian' cycle of erosion.

maturity, less of the original surface would remain on the hilltops, and the valleys would be much larger. River long profiles would tend towards concavity, grading to a stable base level. By old age the hill areas would have been worn down and the landscape would be characterised by extensive plains (peneplains). Nowadays these concepts are seen as hopelessly simplistic, but they dominated research in geomorphology during the first half of the twentieth century. They are based on unrealistic concepts of tectonics and the effects of climatic changes, and do not take into account a true understanding of processes. However, some aspects of the concepts are valid: base level, and the recognition of polygenetic landscapes.

3.3.1 Evidence of past landscapes

Landscapes preserve evidence of past conditions, in patterns, surface form and sediments. We have already seen how drainage patterns can reflect long-term geomorphic development. Surface form can reflect changes in erosional regime. Unless related to local rock resistance, a slope convexity would suggest an acceleration of incision into a higher, more gently sloping, older, more stable landscape. This principle is applicable to relatively active landscapes and is particularly the case for upland dissected plateaux (Figure 3.11A). Such gentle upland surfaces cross-cut the underlying structure, so cannot simply be a response to a near-horizontal resistant bedrock. Such plateau surfaces are often cut into by steeper, younger valley systems, and are therefore interpreted as erosion surfaces, remnants of uplifted ancient landscapes that pre-date the dissection. They are particularly common cross-cutting the structures of the ancient

Figure 3.11 Polycyclic landscapes. A. Erosion (planation) surfaces cut across the rocks of the Massif Centrale, Tarn Valley, France. Note the even skyline of the plateau surface, an erosion surface probably formed during the Late Tertiary that cuts across the geological structures. Note also the deeply dissected Tarn valley (running from left to right across the centre of the photo), which probably developed largely during the Pleistocene. The spur tops in the middle ground probably represent an intermediate stage in the development of this landscape. B. River terraces (*see also* Section 4.3.5), Dane valley, Cheshire, England. The river terrace (at the right of the photo) represents an earlier valley floor formed during the late Holocene. It is composed of gravels at the base, overlain by silts. The river cut through its valley floor, leaving it as a terrace, then, by lateral migration over the last 200 years or so, formed a new floodplain (to the left of the photo, banked against the base of the terrace) below the level of the former valley floor.

Caledonian and Hercynian mountain belts in Europe and in the Appalachians of the USA.

Upland **erosion surfaces** were the focus of much study during the middle years of the twentieth century, but that study was rather inconclusive. The origins of such surfaces are uncertain. Some may be exhumed unconformities where erosion has re-exposed an ancient surface from the geological record. This seems to be the case in parts of the Appalachians and on the dip slope of some Chalk escarpments around the London basin, where the weak Eocene rocks have been stripped away to expose the ancient sub-Eocene surface. Some surfaces might have a marine origin, but most are far too extensive, and perhaps too irregular, to be former wave-cut platforms (*see* section 4.6.2). They appear to be former land surfaces, though their precise genesis is open to question. They have been interpreted as former peneplains in the Davisian sense (*see* above), as analogous to **pediplains** (coalesced pediments: rock surfaces, characteristic of modern arid environments: *See* Section 4.2.1.2), and particularly as **etchplains** (exhumed rock surfaces developed under a deep-weathering regime in a tropical or sub-tropical climate, during the Tertiary). This last interpretation is favoured by most German geomorphologists, substantiated by the presence in places of remnants of tropical soils. What is clear, however, is that such erosion surfaces do represent old land surfaces into which the younger valley system is incised. They are particularly well developed on the Palaeozoic rocks of older (Caledonian and Hercynian) mountain systems, where in many cases 'staircases' of erosion surfaces have been identified. Pulsed epeirogenic uplift (*see* Sections 2.3, 3.1.1) leading to pulsed rejuvenation may have been the cause. In any case they represent relict landscapes, formed during Tertiary time. During the Pleistocene they have undergone dissection; the modern valleys are cut into and below the plateau landscapes of the erosion surfaces. This phenomenon of Pleistocene incision into late Tertiary landscapes is also apparent in areas of 'Alpine'/ Tertiary tectonics. For example, in Spain many areas of late Tertiary internal drainage were dissected during the Pleistocene by the establishment of through drainage.

During the incisional development of the landscape, traces of former valley floors may be preserved as **river terraces** (Figure 3.11B). Their form may allow the reconstruction of former valley floors, and their sediments can yield information on the environments of deposition (*see* Section 4.3.5). In areas affected by ice during the last glaciation, such terraces date only since the melting of those ice bodies. In areas outside the last glacial limits, and particularly outside the maximum glacial limits, river terraces may give valuable evidence for geomorphic evolution through much longer periods of the Pleistocene.

Other sediments, as well as river terrace sediments, can of course tell us a great deal about palaeo-environments. This is particularly true for glacial sediments. It is especially useful when the stratigraphic relationships between sediments of different origins are evident. We shall explore some of the implications in Chapters 4 and 5.

3.4 An example – the regional scale geomorphology of Europe

In this chapter we have reviewed landform development at the regional scale. In order to link the global/continental scale, dealt with in Chapter 2, with the regional scale dealt with in this chapter, we can consider how the regional geomorphology of Europe reflects the plate tectonics context, onto which regional Quaternary climatic patterns and those of the modern environment have been superimposed (Figure 3.12).

The plate tectonics context, operating since the early Tertiary, relates first to Atlantic sea-floor spreading. This accounts for the volcanic activity in Iceland and in the Azores. Aborted rift systems account for the early Tertiary volcanic activity in western Scotland, and Neogene to Quaternary volcanicity in the Auvergne (France) and the Rhine Highlands (Germany). The second aspect of the plate tectonics context relates to the convergence of the African and European plates, culminating in the mid-Tertiary formation of the Alpine fold and nappe system. That activity is ongoing today, as is evidenced by subduction and volcanic activity in Greece and southern Italy. Elsewhere within the Alpine system, **neotectonic activity** operates, whereby **epeirogenic** or post-orogenic uplift, and local folding and faulting continue.

During the Pleistocene ice ages large ice sheets formed over Scandinavia and northern Britain, with a smaller ice cap over the Alps and smaller glaciers in some of the other European mountain ranges. Periglacial climates with permafrost covered the non-glaciated areas of southern Britain and most of western and central Europe. There was extensive loess deposition south of the limits of the Scandinavian ice sheet, extending through Belgium and Germany into eastern Europe. Global sea levels were low, so that the continental shelf around southern Britain was dry land.

Present day climates influence geomorphic processes. Most of western Europe has a humid temperate climate, ranging from mild maritime conditions on the Atlantic seaboard to continental conditions (cold winters, hot summers) in eastern Europe. Snow is important in eastern Europe, with its associated spring snowmelt flooding. The effectiveness of frost, snow and ice increases with altitude, so that the Scandinavian mountains and the Alps and Pyrenees have truly mountain climates. Throughout western and central Europe rain is all-year with a winter maximum in the west and a summer maximum in the east. Heavy cyclonic or convectional storms can lead to slope instability, especially in upland and mountain areas, and to river flooding throughout Europe. 'Natural' conditions would support deciduous forest over most of Europe and a well-developed soil cover, so that surface erosion would be minimal, but mass movements can occur on steeper slopes. Today much of the forest cover has been replaced by agricultural land, some of which is prone to limited soil erosion.

A Mediterranean climate characterises most of southern Europe, with hot, dry summers and autumn to winter rain. Especially in autumn, this can be torrential. In the drier parts of the Mediterranean the climate is semi-arid, with marked soil moisture deficits. In these conditions soils are slow-forming. The 'natural' Mediterranean scrub woodland has in many places been seriously degraded by a long history of overgrazing and land abuse, so that run-off, surface erosion, and flash flooding are common in response to torrential storm rains.

In Box 3.1 and on Figure 3.12 we summarise the main regional geomorphology of Europe, in so far as it reflects the interplay of these three groups of factors.

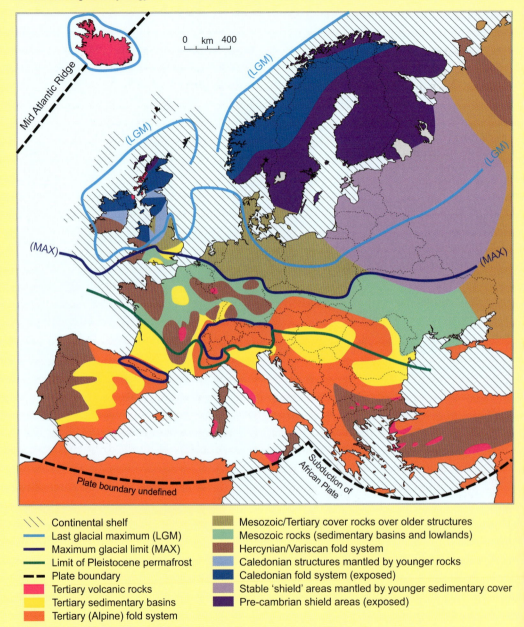

Figure 3.12 Geomorphological regions of Europe, showing the major structural units, and the Pleistocene glacial and permafrost limits.

Box 3.1 Europe: geomorphic regions (see Figure 3.12)

1. The ancient core of the continent

1a) Where the ancient rocks are exposed – the Baltic Shield

This area has been structurally stable since Precambrian times. The ancient metamorphic rocks have been much eroded down to produce subdued relief. This area was repeatedly heavily glaciated by the Scandinavian ice sheet during the Pleistocene, with glacial erosion dominant, resulting in an irregular landscape of bare rock outcrops with intervening lake basins. Glacial depositional topography becomes more important towards the Baltic coast.

A small area in Assynt district and the Outer Hebrides of Scotland is another area of shield terrain, a remnant of another Precambrian shield area with affiliations not with the Baltic Shield but with the Canadian shield.

1b) Where the ancient rocks are mantled by near flat-lying younger sedimentary rocks – the North European Plain.

To the south of the Baltic Shield the structurally stable ancient rocks are mantled by near flat-bedded younger sedimentary rocks to form the North European Plain. Here the relief is very subdued. The landforms are dominated by glacial and fluvio-glacial deposition. Major fluvio-glacial meltwater channels and moraines form the main features of the relief.

2. Two ancient mountain chains (related to ancient plate-tectonic settings, pre-dating the Atlantic Ocean)

2a) The Caledonian system (Norway, Scotland, N Ireland – continues into NE USA and Canada).

This system includes metamorphics, igneous intrusions, volcanics and folded sedimentary rocks. They are mostly structurally complex, but in some places are covered by uniclinal, or gently folded Upper Palaeozoic sedimentary rocks. The former mountains have been much eroded down to form plateaux, then later uplifted and deeply dissected. These mountains were heavily glaciated during the Pleistocene, producing glaciated mountain scenery. Glaciers are still present in Norway.

2b) The Hercynian system (central Europe, central and W France, parts of Spain, SW England, S Wales, SW Ireland – continues into the Appalachians).

This system also comprises metamorphic rocks, igneous intrusions and volcanic rocks, together with folded Upper Palaeozoic sedimentary rocks. Much eroded down to form plateaux, then uplifted and deeply dissected. These areas currently form separate upland blocks rather than a continuous system. A very few small areas were glaciated during the Pleistocene, but periglacial processes predominated.

3. The Tertiary 'Alpine' mountain chains

3a) The Alps

The core areas of this large mountain system are the metamorphic rocks, including some incorporated from older Hercynian structures of the central area of the French, Italian, and Swiss Pennine Alps. The rocks are deformed into huge complex overfolds (nappes) which have been thrust towards the north. Younger nappes of metamorphic rocks have been thrust onto the Alps of the Bernese Oberland, north of which are the folded Cretaceous limestones of the pre-Alps. North of these lies the detrital Swiss midland plain, then the thrust-forward, folded Jurassic limestones of the Jura Mountains. To the south-east of the Pennine Alps are the limestones of the Dolomites and the Dinaric Alps, which continue south-east through the Balkans into Greece. During the Pleistocene, repeated ice caps formed over the central Alps, extending west to the Jura and into the Rhone valley near Lyon, and north-east onto the German Alpine foreland. Small glaciers persist today within the highest mountain areas.

3b) Other 'Alpine' mountain systems (The Apennines, Pyrenees, Betic and Iberian systems)

The Apennines extend south-east from the French Maritime Alps, forming the backbone of Italy. In contrast to the main Alpine system with its arcuate alignment, the Pyrenees represent an earlier east–west compressional trend that continues into the folded limestone Alps of Provence. Further south in Spain are a series of 'Alpine' ranges, the Iberian and Central Systems of lower elevation, but in the far south are the high Betic Ranges. These comprise a metamorphic core and a northern belt of folded limestones in the sub-Betic ranges. The Betics continue into the Rif of Morocco. Of these secondary ranges, the Betic Ranges supported tiny Pleistocene glaciers, but only the Pyrenees supported glaciers of any size. The Pyrenees are the only ranges to support small glaciers at present.

4. The intervening areas: scarplands and lowland basins

Between the uplifted fault blocks of Hercynian structures, between these and Alpine structures, and within and between Alpine structures are zones of the crust that are relatively depressed. These zones are blanketed by sedimentary rocks younger than the surrounding uplands or mountains. Some of these are of relatively shallow depth, where the younger rocks simply bury the older structures (e.g. a buried platform of Hercynian structures that extends from the English Midlands under London, south-eastwards into Belgium). Others are distinct sedimentary basins of Mesozoic and Tertiary age (e.g. the Paris Basin), or of Tertiary age within or between Alpine structures (e.g. the Ebro Basin; the Saone Basin; small sedimentary basins within the Betic Cordillera of southern Spain). Those basins within the Alpine zone continue to be subject to neotectonic activity.

These areas are generally lowlands, but where a stack of sedimentary rocks of differing erosional resistance is involved, scarpland topography is common. Most of these areas were not glaciated during the Pleistocene, except the English Midlands and East Anglia, and the Netherlands, where the area merges into the north European plain. However, in western Europe these areas underwent periglacial processes. Only in Iberia did the sedimentary basins escape periglacial activity.

4 Local-scale geomorphology – process systems and landforms

The local scale is the basic scale of geomorphology. It is the scale at which we recognise landforms, e.g.river reaches, hillslopes, cliffed coasts. It is also the scale at which we understand the processes that create the landforms. At this scale we are dealing almost exclusively with externally-driven geomorphic processes. These processes are parts of the 'sediment cascade' (*see* Figure 1.7) whereby bedrock is transformed by weathering, then transported, with the aid of gravity under one of the process regimes described below, to either temporary or more permanent deposition. The resultant landforms may therefore be dominantly erosional, composite or depositional.

The only real exceptions to the dominance of externally driven processes are the direct results of volcanic activity – fresh volcanic cones, lava flows; or of tectonic activity – fresh fault scarps (*see* Section 3.1.2). Even then, in a relatively short time these landforms undergo modification by weathering and other surface processes.

4.1 Weathering systems

Weathering is the breakdown of raw rock by mechanical or chemical means into material that can be incorporated into soil or transported away by one of the geomorphic transport processes described below. Alternatively the weathered rock may remain *in situ* as '**regolith**', and may later be subject to erosion and fed into the sediment cascade (Figure 1.7)

Weathering is an essential prerequisite for the operation of the sediment cascade.

4.1.1 Mechanical weathering

Rocks are rigid elastic solids and respond (in a small way) to ambient pressure. Igneous rocks, especially **plutonic** rocks, and metamorphic rocks are formed within the crust at pressures much higher than atmospheric pressure. Even sedimentary rocks have been subject to loading by overlying rocks. Below glaciers and ice sheets, rocks are also subject to loading. When exposed to the much lower atmospheric pressure, rocks expand elastically. In layered rocks this may simply enlarge existing bedding planes, but in massive rocks (e.g. granites) this expansion may result in cracking parallel to the surface of the ground (Figure 4.1A). The cracks then provide access for water, which may further weather the rock. This process is described as **pressure release** or offloading jointing.

Rocks are also subject to cracking as the result of thermal expansion. The efficacy of this mechanism was doubted when it could not be duplicated in experimental ovens; however, more recent work in deserts, by Amit, Gerson and McFadden among others, has demonstrated very strong temperature gradients between the sunny and the shaded sides of rocks, sufficient to generate cracking. Desert surfaces composed of angular stones (**desert pavements**) are formed in this way (Figure 4.1B). In saline environments, for

Figure 4.1 Photos of mechanical weathering phenomena. **A.** Pressure release (offloading) joints in granite, Yosemite Valley, California, USA. These joints are parallel with the valley sides, reflecting offloading on deglaciation. **B.** Desert pavement surface, Sinai Desert, Egypt. Note how rocks are cracking to form the angular clasts of the pavement. Note also the silt between the clasts on the pavement surface. **C.** Frost-shattered slates, Rocky Mountains, Canada. **D.** Patterned ground (stone stripes) near the summit of Mont Ventoux, Provence, France. These stripes were formed during the Pleistocene when permafrost was present. **E.** Involutions and other features related to the presence of Pleistocene permafrost in alluvial gravels and silts of an alluvial fan exposed at Doniford, Somerset, England.

example near salt flats in deserts, the process may be accentuated by **salt weathering**. After wetting of the rock surfaces by rain or dew, salts are precipitated within cracks in the rock, crystal growth further cracking the rock.

Another important mechanism for fracturing rock is freeze-thaw weathering. Water expands on freezing, so when water that has penetrated into cracks in rocks freezes, it exerts sufficient pressure to fracture the rock. This process is, of course, important in high latitude and high mountain areas, subject to sufficiently low freezing temperatures and to frequent freeze-thaw cycles. It is one of the main mechanisms for supplying debris from rock surfaces onto scree slopes below (Figure 4.1C).

Another aspect of freezing and thawing is the development of **patterned ground**. In areas of permafrost, refreezing of the **active layer** (*see* Section 4.2.1.3) creates stresses in the soil because of the different thermal properties of stones and finer material. This results in a sorting of the material so that on flat ground the stones form polygons; on gently sloping ground these elongate into garlands, and on steep slopes form stone stripes running downslope. At depth within the soil these are represented by re-orientated stones forming **involutions**. At greater depths, ice wedges may form where water trapped by freezing of the active layer is forced into the upper layers of the permafrost. Relict patterned ground, involutions and fossil ice wedges (**ice-wedge casts**) are indications of cryoturbation and the former presence of Pleistocene permafrost in temperate latitudes (Figure 4.1D, E).

There is one other important mechanism that forms or enlarges cracks in rocks; that is through the action of tree roots.

4.1.2 Chemical weathering

Chemical reactions in soil and rock depend on the acidity of water passing through the material, on temperature, on the availability of oxygen and on the susceptibility of the material itself to chemical change. Rainwater is mildly acidic. Organic soils release humic acids, so that water that has passed through these soils is acidic. In arid areas soils tend to be alkaline.

A range of chemical processes may be involved in the weathering of rock. The simplest of these is **solution**. We have already seen how calcium carbonate ($CaCO_3$), the main constituent of limestone, is soluble in weak acids, producing 'karst' solutional features in limestone terrain (Section 3.1.3). $CaCO_3$ can also be a major cement binding the grains together in sandstones, which on solution renders the sandstone into loose sand. Another important process is the absorption of water into the crystal lattice of some minerals (e.g. the transformation of anhydrite, $CaSO_4$, into gypsum, $CaSO_4.2H_2O$) by **hydration** or the reverse process, **dehydration**. Another important set of reactions involves the addition or loss of oxygen to or from a mineral: **oxidation** and **reduction** respectively. The simplest example is the transformation of a metal into an oxide, but much more important in the context of weathering and soils are the transformations between the two families of iron oxides: oxidation to ferric oxides (the red/brown iron oxides: Fe_2O_3) and reduction to ferrous oxides (the black/grey iron oxides: Fe_3O_4). Oxidation takes place in aerated environments, whereas reduction takes place in oxygen-poor, often water-saturated environments. Bacterial action accelerates these processes. **Hydrolysis** is a more complex chemical reaction involving

Table 4.1

A. Primary igneous rock-forming minerals and mineral groups ranked by their chemical stability in weathering environments.

Quartz	**Si O$_2$**
Muscovite Mica	**K Al (Al Si3) O$_{10}$ (OH)$_2$**
Orthoclase Felspar	**K Al Si$_3$ O$_8$**
Plagioclase Felspar	**Na-Ca Al Si$_3$ O$_8$**
Biotite Mica	**K (Mg Fe)$_3$ (Al Si$_3$) O$_{10}$ (OH)$_2$**
Hornblende	**Complex silicate of Mg Fe Ca Al**
Augite	**Complex silicate of Mg Fe Ca Al**
Olivine	**(Mg Fe)$_2$ Si O$_4$**

Red indicates important constituents of acid rocks (eg. **granite**); **blue** indicates minerals that occur across the range of acid to intermediate to basic rocks; **purple** indicates minerals that might occur in some acid rocks, and also occur in intermediate rocks (eg. syenite, dacite, andesite); **green** indicates minerals that occur in basic or ultrabasic rocks (eg. gabbro, **basalt**).

Quartz is almost chemically inert, but may break down mechanically to sand. The other minerals break down chemically to clay minerals, the ferromagnesian minerals also to metal oxides (particularly iron oxides).

B. Minerals in sedimentary rocks, include those that have been derived directly from igneous rocks (quartz, muscovite, some feldspar), the weathering products of primary minerals – particularly clay minerals and iron oxides, plus chemical or biochemical precipitates (especially **calcite**, Ca CO$_3$).

C. Metamorphic rocks may include an enormous variety of minerals, derived from their igneous or sedimentary rock parent, plus in high grade metamorphic rocks a suite of complex silicates found only in metamorphic rocks.

the exchange of ions, usually in mildly acidic conditions. This is the main process involved in the weathering of the primary rock-forming minerals of igneous rocks. A final weathering process that is perhaps more important in soils than in rock weathering is **chelation**, the mobilisation of metal ions from clay minerals, allowing leaching to take place.

Different rock-forming minerals have differing susceptibility to chemical weathering.

Minerals characteristic of igneous (and high-grade metamorphic) rocks were formed in environments very different from those at the Earth's surface, so tend to be susceptible to chemical change. However, their susceptibility differs between mineral groups (Table 4.1). Quartz (SiO$_2$) is virtually inert, except that it becomes soluble in very alkaline environments. The feldspars (complex potassium/sodium/calcium aluminium silicates) and muscovite mica break down by hydrolysis into a complex family of clay minerals (complex platy hydrated aluminium silicates), though muscovite is less susceptible to chemical change. The ferromagnesian groups of minerals (complex ferromagnesium silicates), biotite mica, the hornblende, augite and olivine groups, are increasingly susceptible to chemical change, especially by hydrolysis to produce oxides and clay minerals. The minerals present in sedimentary rocks include detrital minerals (quartz, muscovite, clay minerals) that have already been through a weathering cycle during their formation. The more susceptible minerals include cements and chemical/biochemical precipitates or evaporites (e.g. $CaCO_3$, iron oxides, gypsum), which have often been formed in environments somewhat different from surface weathering environments, and so are susceptible to chemical change.

The implication of differing mineral susceptibility to chemical change is that the vulnerable minerals are attacked first, weakening the fabric of the rock, allowing an acceleration of weathering processes as a whole. Chemical weathering also affects the physical properties of the rock, often making it mechanically weaker, less coherent and less dense, hence more easily erodible. In the weathering of granite, for example, original vertical cracking and sub-horizontal pressure-release jointing allow selective chemical weathering to penetrate the rock along the joints, transforming the weathered granite into **grus** (a mixture of sand and clay) surrounding unweathered **corestones** (Figure 4.2A). If the grus is washed away the corestones remain as **tors** (Figure 4.2B). Similar features in tropical Africa are known as **kopjes** (Figure 4.2C).

Figure 4.2 Granite weathering. **A.** Corestones, surrounded by 'gruss', exposed in a roadcut, Capetown, South Africa. **B.** Tors, Dartmoor, Devon, England. Note the dominance of (pressure release) joints parallel with the ground surface, and also the vertical joints. **C.** A kopje, Northern Cape Province, South Africa.

4.1.3 Influence of climate on weathering regimes

Because both mechanical and chemical weathering depend on temperature and moisture conditions, there is a climatic control of weathering regime. In arctic and alpine regions mechanical weathering by freeze-thaw is dominant and chemical weathering is weak. In humid temperate regions all processes are moderately effective. In arid regions weathering overall is weak, but mechanical weathering is dominant. Arid regions, though, have distinct chemical regimes (*see* below). In the humid tropics chemical weathering is intense, resulting in great thicknesses of a deep weathering mantle. Many areas in the tropics and sub-tropics, especially on ancient stable land surfaces, such as those on shield areas (e.g. the Yilgarn block in Western Australia), preserve a weathering profile of great thickness (up to 50 m deep) (Figure 4.3A), that has developed over timescales extending back into the Tertiary. At the base of such profiles is partially weathered bedrock overlain by a deep red horizon, pallid then mottled zones, finally capped by

Figure 4.3 Soils and weathering profiles. **A.** Deep weathering profile, near Cue, Western Australia. At the base are unweathered corestones of the granitic bedrock, above which is pale gruss of the 'pallid' zone; the section is capped by ferricrete (formerly known as laterite). The deep weathering profile developed during the Tertiary under more humid conditions than characterise today's arid climate. **B.** Podzol, developed on a late Pleistocene gravel terrace, Howgill Fells, Cumbria, England. Note the bleached horizon immediately below the dark humus horizon. Below is the iron-rich B horizon. **C.** Desert soil, Dixie Valley, Nevada, USA, developed on a Pleistocene alluvial fan surface. Note the mature desert pavement at the ground surface. Immediately below that is the vesicular Av horizon composed of desert dust containing soluble salts. Below that is the dark red-brown clay-enriched Bt horizon, below which is the pale carbonate-rich Bk horizon.

a **ferricrete** or **silcrete duricrust**. Rarely is the bedrock exposed at the surface. More often the surface relief is formed by 'breakaways', outcrops of the duricrust (*see* below).

Weathering processes convert less stable rock and mineral materials into more stable products: rock fragments, quartz sands and clay minerals. The fate of the soluble products depends on hydrology and climate. In humid areas in free-draining sites, leaching of the soluble products through the soil profile takes place. These products may be carried away as the dissolved load of streams, eventually to the oceans. Locally, either in the soil profile or elsewhere, if the chemical environment is suitable they may be precipitated. The classic soil profile of cool humid areas with acid soils is the **podzol** (Figure 4.3B), where all carbonate is completely leached from the soil, but iron leached down the profile accumulates at depth, its status (oxidised or reduced) depending on the drainage of the soil.

In dry regions, leaching is much less important. Iron compounds (ferric iron) remain in the upper part of the soil profile, giving it a reddish colour. Carbonates tend to be leached from the upper part of the profile but are precipitated and accumulate lower down (Figure 4.3C). On later wetting and drying and exposure to the atmosphere, this **pedogenic** carbonate may be transformed into a resistant caprock, a **calcrete**. This is a form of **duricrust** (Figure 4.4A), characteristic of

Figure 4.4 Weathering-related phenomena of dry regions. **A.** Calcrete developed on Pleistocene alluvial fan sediments, near Murcia, south-east Spain. This pedogenic calcrete probably developed from carbonate accumulation within a soil profile that was later exposed and became indurated. **B.** A case-hardened rock surface protecting honeycomb weathering (tafoni), near Hatta, UAE. **C.** The end of a fluvial system in the Nevada desert, USA, evaporite salt flats.

drylands. It is important in two ways: firstly, in the study of dryland geomorphology, the properties of calcrete may allow the relative dating of the age of the surface (*see* Chapter 5). Linked to the calcium ions within the calcrete may be uranium ions, which would allow precise dating of the time of crystallisation by the uranium/thorium method (*see* Chapter 5). Secondly, a duricrusted surface is erosionally resistant and also alters the infiltration characteristics of the surface. Calcrete, characteristic of drylands, is not the only form of duricrust; **gypcrete** may also occur in very arid areas, while **silcrete** and **ferricrete** (*see* above; formerly known as **laterite**), occur in the more active weathering environments of the seasonal wet/dry tropics.

Another weathering phenomenon associated with dry conditions is termed **case hardening**. Salts mobilised by solution in near-surface zones of a rock are precipitated as the moisture is evaporated from the surface of the rock. The result is an indurated 'case hardened' outer layer of the rock, which may allow weathering of the softer material underneath to form **tafoni** or honeycomb weathering (Figure 4.4B).

The final fate of solutes in arid areas, especially those entrained by overland flow, may be deposition in an ephemeral lake, which on evaporation to dryness forms a salt pan or **playa** (Figure 4.4C). **Evaporite** salts may form concentric rings around the margins of the playa, ranging from calcite through gypsum to the most soluble, halite or rock salt (NaCl) in the centre.

4.2 Slope systems

Slopes of one sort or another form the vast majority of the landscape; even flat surfaces can be seen as zones of zero slope. However, when we refer to slope processes we really mean hillslope processes. In most areas hillslopes are organised into drainage basins (*see* Chapter 3). Hillslopes thus not only provide the main sources of water to river systems through the mechanisms of the hydrological cycle (*see* Chapter 2), but are also the main sources of sediment. Following bedrock weathering, hillslope processes are the first zone of the sediment cascade. Sediment is mobilised on the hillslopes by a range of processes and moved downslope. Some of this sediment may be deposited further downslope, and some may reach the river channels. Consequently, the form of the slope is the result of hillslope processes.

4.2.1 Slope processes

Four groups of hillslope processes can be recognised, differing according to the nature of the material involved and its behaviour under erosion or deformation: scree processes, overland flow processes, creep and flowage processes, and landsliding.

4.2.1.1 Rockfall and scree processes

These processes involve rock fragments, released by mechanical weathering, falling from an exposed rock face. Large-scale rockfalls can be a major slope process in steep mountain terrain, often leading to a jumbled mass of broken rocks below. At a more limited scale, persistent rockfalls may yield rock fragments that accumulate on the slope below, forming a scree. The scree is deposited at the **angle of rest** of the material (generally around 30°) (Figure 4.5A). Screes often show some downslope sorting of material by particle size, with the larger **clasts** (having more

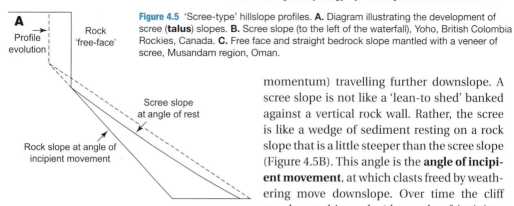

Figure 4.5 'Scree-type' hillslope profiles. **A.** Diagram illustrating the development of scree (**talus**) slopes. **B.** Scree slope (to the left of the waterfall), Yoho, British Colombia Rockies, Canada. **C.** Free face and straight bedrock slope mantled with a veneer of scree, Musandam region, Oman.

momentum) travelling further downslope. A scree slope is not like a 'lean-to shed' banked against a vertical rock wall. Rather, the scree is like a wedge of sediment resting on a rock slope that is a little steeper than the scree slope (Figure 4.5B). This angle is the **angle of incipient movement**, at which clasts freed by weathering move downslope. Over time the cliff recedes at this angle (the angle of incipient movement) and the rock slope becomes buried by scree, which comes to rest at the slightly less steep angle of rest. Screes are common in hill and mountain areas, where present or past processes have created bare rock cliffs, especially where mechanical weathering is or was active. Screes occur in arid areas (Figure 4.5C) but are most common in arctic and alpine areas. In these areas the simple scree processes may be complicated by the addition of material from snow avalanches. Fossil screes are also common in areas where freeze-thaw weathering is now limited, but was active during Pleistocene cold phases.

4.2.1.2 Overland flow

A second group of hillslope processes relates to erosion by overland flow. The simple theory, devised by Robert Horton, envisages rain intensity in excess of infiltration capacity (*see* Section 1.4.2) to pond water on the surface, then for it to run off downslope. Initially flow is by laminar flow, which exerts no stress on the underlying surface, creating in the divide area a **belt of no erosion** (Figure 4.6A). With increasing distance from the divide, turbulent flow sets in, initially entraining any loose particles

Figure 4.6 (Opposite) Features produced by overland flow erosion. **A.** The 'belt of no erosion' in the divide area on badland slopes, Alberta, Canada. Note that the rills start a little way below the divide. **B.** Gullied badland terrain, northern Provence, France. Note the smooth 'belt of no erosion' on the divides, the steep straight slopes of the belt of rill erosion, the much gentler gradient of the main gully floors. **C.** Piped badlands near Sorbas, Almeria, south-east Spain. The cavities on the left are entry points to an extensive network of underground pipes. **D.** A complex slope profile suggesting response of geomorphic systems to late Quaternary climatic change, Capitol Reef National Park, Utah, USA. The upper part of the slope is dominated by a free face of bare rock and a (now degraded) scree slope, probably active during colder phases in the Pleistocene. The lower part of the former scree slope is being eroded by badland development, related to today's semi-arid environment. Note the characteristic development of badland morphology and the development of mini-pediments at the base of the slope, caused by the recession of the steeper rilled slopes. Note the pediment angle (*see* Figure 4.10C). **E.** Large-scale pediment development below badland slopes, near Hanksville, Utah, USA. **F.** Combinations of hillslope gullying and valley-floor gullying (arroyo-type), near Drumheller, Alberta, Canada.

on the surface. Then as the flow becomes concentrated into shallow rills it removes weathered material. The rills are essentially at the same gradient as the surface itself, and together they form **zones of sheet and rill erosion** (Figure 4.6A, B). As run-off further increases downslope and erosional stresses become sufficient to incise into the underlying material, a **gully** channel is cut (**zone of gully erosion**: Figure 4.6B), and because flow is now concentrated and depths are much greater, the channel develops at a gradient markedly less steep than that of the hillslope.

This (Hortonian) model has shortcomings for two reasons. First, other processes operate in rapidly eroding gullied terrain. Shallow mass movements (mudslides) modify the erosional processes. Run-off can be generated if

infiltration capacity is zero because of soil saturation (**saturation overland flow**), and though this may not be so important erosionally, it is important hydrologically. In many rapidly eroding areas, especially deeply dissected bare **badland** areas, water may penetrate into the ground through cracks and sustain subsurface turbulent flow capable of **tunnel or pipe erosion** (Figure 4.6C). In many badland areas, piping may be as important as surface erosion. A second drawback of the Horton model is that it is essentially a static model and cannot take into account the progressive development of eroding slopes. The topography only holds true if the gully channels continue to incise, keeping pace with erosion on the slopes. If, on the other hand, the gradient of the gully channels is controlled by a local base level, the channels simply transport the sediment supplied by the eroding hillslopes and cannot incise. In that case, the eroding slopes undergo parallel retreat, creating small **pediments** at the slope base (Figure 4.6D), ultimately resulting in small erosional hills above extensive pediments (Figure 4.6E). Pediments (*see* below, Section 4.2.2) can also form in more resistant rocks, and are characteristic landforms in dry regions. They act essentially as surfaces of sediment transport by sheet flows. They may carry a veneer of sediment, but the main process operating on the surfaces is mechanical weathering. In deserts they may be veneered by desert pavements (*see* above, Section 4.1.1).

Despite its shortcomings, the Horton model does provide a basis for understanding erosional terrain. It is most applicable in bare areas or areas of thin vegetation on soils with low infiltration capacities that are prone to high intensity rainstorms. In the short term, gullying in badland areas is an unstable process, reinforced by positive feedback (*see* Section 1.5). Gully channel incision tends to increase the drainage area to the main gullies, thus increasing run-off and incision rate. The process only slows when a longer term negative feedback begins to operate. Competition between neighbouring gullies prevents incision from continuing to increase drainage areas. From then on erosion simply lowers the divide areas, progressively reducing the rate of erosion.

Badland or gullied terrain, characterised by a high **drainage density** (*see* also Figure 3.2B) is a common natural terrain in semi-arid areas. Such terrain also occurs in other climatic zones on sea cliffs, river cliffs and landslide scars cut in soft materials. This type of terrain can also be produced by human activity, by overgrazing, compaction of the soil, or by plough lines creating ready-made rills, and so may be an indication of past or present human-induced soil erosion (*see* Chapter 6).

A related aspect of overland slope erosion, equally applicable to human-induced soil erosion, is the down-system fate of the sediments. Some sediment may be stored within the slope system and some fed down-system into the river system. A wave of hillslope erosion or soil erosion may have a dramatic effect on river behaviour and morphology (*see* below, Section 4.3.4). In extreme cases, especially in semi-arid areas, the sediment may totally overload headwater valleys and cause valley filling. Subsequently this valley fill may become dissected by valley-floor gullies (e.g. the **arroyos** of the American west). Severely eroded (badland) terrain in such environments may show combinations of hillslope and valley-floor gullying (Figure 4.6F).

4.2.1.3 Mass movement processes (excluding landslides)

The third group of slope processes (excluding landslides; *see* below) are those mass movement processes that involve deformation of weak or unconsolidated **regolith** material, the weathering mantle or soil. The behaviour of regolith material depends on its moisture content. Under dry conditions and light loading it may deform by cracking/fracture, but with increased moisture content its **plastic limits** are exceeded, and it will deform under loading by an internal rearrangement of its shape, i.e., by plastic deformation or flowage. With further increased water content the behaviour crosses another threshold, the **liquid limit**, beyond which the material behaves as a fluid and drains more rapidly under its own weight.

The rate at which slope processes by mass movement operate may range from the imperceptible to the catastrophic, and may be widespread or highly localised. There have been a number of classifications of mass movements, based on the nature of the material involved (rock fragments, mixed debris, soils), water content, and the nature and speed of the movement, but here, for simplicity's sake, we will treat the most common in three groups: creep, solifluction and debris flows.

The most ubiquitous, but almost imperceptible, process is **soil creep**. On wetting or freezing the soil swells, expanding upwards orthogonal to the surface. On drying or thawing the soil shrinks and sinks, but with a gravity-driven downslope component, resulting in the net downslope movement of the upper part of the soil layer. This process is difficult to observe, except when ice needles are involved, but evidence for its operation

Figure 4.7 Forms produced by (slow) mass movement processes. **A.** Evidence for surface creep expressed by downslope curvature of weathered bedrock fragments, Clwydian Hills, North Wales. **B.** Hillslope terracettes, Ellen valley, Cumbria, England. **C.** Evidence for sustained Pleistocene hillslope solifluction, a hillslope blanketed by thick 'head deposits', north Cornwall, England. These deposits comprise angular (local) stone fragments, derived from higher up the hillslope by freeze-thaw activity under permafrost conditions, then transported downslope by solifluction, to leave a muddy-stony deposit in which the clasts are aligned downslope.

is widespread in the form of bent trees on hillslopes, re-orientated stones within the upper part of a soil profile (Figure 4.7A), or **terracettes** on steep grass-covered hillslopes (Figure 4.7B). In Britain these features are sometimes referred to as 'sheep tracks', but they form in the absence of grazing animals, and are likely to be the result of soil slip or flowage on a microscopic scale.

A similar, but larger-scale and faster process, **solifluction**, is characteristic of permafrost environments. During the summer the upper parts of the regolith profile (the **active layer**) thaw out, but drainage is prevented by the underlying impermeable frozen layer, the **permafrost**. Shallow lobes of material move relatively slowly, but measurably, downslope, and in some areas whole hillslopes can be blanketed with solifluction lobes. Eventually these may lose their topographic expression, resulting in a smooth hillslope (Figure 2.5A). This is a major process affecting arctic hillslopes. It was the major slope process during the Pleistocene cold phases, affecting hillslopes in what are now temperate areas. Evidence is preserved partly in hillslope form, and especially in 'head deposits', which blanket many hillslopes in these areas (Figure 4.7C). These are deposits of angular stones set in a finer matrix, with the major planes of the stones roughly orientated downslope.

A faster flowage process, and one that may be more or less channelised, is **debris flow** (Figure 4.8A). The material, usually a mixture of all particle sizes, may be derived from shallow slope failures or be entrained from within-gully debris. It is usually mobilised during heavy rainstorms or during snowmelt. The water content determines how the flow behaves. A relatively low water content (below the liquid limit) will result in a cohesive debris flow, which, especially if the clay content is high, will deform by internal compression and shearing. It will have the strength to support large clasts, boulders, on the surface. It produces a distinctive lobate depositional topography (Figure 4.8A), with levees left behind the lobe on either side of the flow path. Its sedimentology is also distinctive, with an internal structure characterised by

Figure 4.8 Debris flows. **A.** Debris flow on the surface of a debris cone, Howgill Fells Cumbria, England. Note lobe and levee topography of the debris flow. **B.** Bouldery debris-flow deposits on alluvial fans at Zzyzx, California, USA. Note the 'push fabric' of the boulders in the upper part of the photo.

clasts supported within the matrix, a concentration of large clasts at the front and on the surface of the lobe, and a clast alignment across the front of the flow, a compressional fabric (Figure 4.8B), and aligned parallel with the flow along the levees. As water content increases, the internal strength begins to break down, flow velocity accelerates, and the flow regime is transitional between cohesive and fluid flow, known as 'hyperconcentrated' flow. The lobe and levee topography will be less pronounced and the internal structure may be more diffuse. The internal clasts, still matrix-supported, may be aligned with the flow, but on deposition the wet matrix may drain, leaving an irregular stony fabric. Though not relevant to hillslope processes, similar processes can operate under water, to create sub-aqueous debris flows, though generally these would show sediment properties related to more fluid conditions. Debris flows also occur on glacial margins, and can involve the debris from landslides. Very rapid and incoherent debris flows in mountain areas are known as debris avalanches, and the particular type of debris flows associated with volcanic eruptions are known as **lahars**. All debris flows are potentially hazardous, particularly lahars.

4.2.1.4 Landslides

Landslides differ from other mass movements in that they involve the development of a failure surface on which a large mass of material slides downslope. The failure surface, developed internally, may be arcuate or planar (Figure 4.9A,B,C). As the material

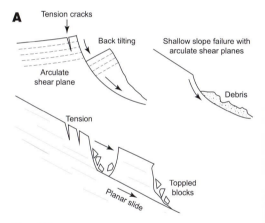

Figure 4.9 Landslides. **A.** Planar and arcuate shear planes. **B.** Planar slide on the East Devon coast, England. The rocks dip gently seawards. Marine erosion has destabilised the slope, activating a planar landslide surface in the Triassic marls at the base of the cliff. The stronger Cretaceous sandstones above have slid along this shear plane, leaving a chasm behind. **C.** An arcuate landslide developed above a rapidly incising river, Sorbas area, south-east Spain. Note the curved landslide scar – the centre of the failed mass has been levelled for agricultural purposes. **D.** Landslip on the Cotswold Escarpment at Birdlip, Gloucestershire, England. The landslip scar is hidden in the trees to the left of the photo; the landslide debris forms the hummocky ground in the centre of the photo.

This landslip probably occurred during thawing of the permafrost at the end of the Pleistocene. **E.** Small landslide feeding debris flows, Southern Alps, South Island, New Zealand. Note the alluvial fan to the left of the debris cones.

moves away it leaves a landslide scar behind, on which other slope processes – rilling, debris flows, etc. – may develop. As the mobile mass settles in an accumulation zone it may remain more or less intact or disintegrate to form ill-drained hummocky terrain (Figure 4.9D) on which other processes, (e.g. debris flows) may develop. Indeed, shallow landslides are a major initiation point for hillslope debris flows (Figure 4.9E).

Factors which influence the occurrence of landslides are:

1) Excessively steep slopes, especially if they are undermined by erosion at the base, in situations such as river or sea cliffs (Figure 4.9B).
2) Geological structure: an alternation of weak and strong rocks is conducive to the formation of shear planes within the weaker rocks. Downslope dips to the strata may foster landslide occurrence (Figure 4.9C).
3) Groundwater conditions. Concentrations of groundwater, or steep gradients to the water table, may foster landslides.

The triggers for landslide events include:

1) Basal erosion and the removal of basal support to the hillslope.
2) Heavy or long duration rainstorms with increased percolation to the water table, causing saturation within the rocks, especially when concentrated along particularly weak geological horizons.
3) Melting of permafrost.
4) Earthquake shocks.

Often more than one of these triggers may be involved. For example, the first typhoon to hit areas in south-east Asia after a major earthquake may trigger an excessive number of landslides.

Landslide zones may be active or inactive.

For example, many of the landslide zones recognisable in Britain are currently inactive and date from the end of the last glaciation, or from the melting of Pleistocene permafrost (Figure 4.9D). However, landsliding is a major process on coastal cliffs, and is particularly important in young mountain areas, where it may pose a serious hazard to human activity (*see* Chapter 6).

4.2.2 *Climatic influence on slope processes and slope profile morphology*

Because many of the drivers of slope processes are climatically controlled, the processes themselves and the resulting morphologies have climatic expression. In arid landscapes erosion by overland flow is much more important than mass movement, whereas the reverse is true for humid soil-covered landscapes. In arctic areas the presence of permafrost influences slope processes, and of course many of the present slope forms in temperate latitudes are fossil forms inherited from Pleistocene periglacial conditions.

The processes also have an influence on slope profile morphology. There have been a number of slope profile models, some of them quite elaborate. It is probably best here to present a simple model (Figure 4.10) on which four facets are recognised, each of which may be produced by a particular set of processes. Mass movement by creep and flowage would be dominant on the upper convexity; mechanical weathering and rockfall would be dominant on the exposed bare rock free-face segment; scree processes would be dominant on the straight slope segment; and overland flow processes would be dominant on the basal concavity. In theory, combinations of these facets or emphasis on particular facets would be

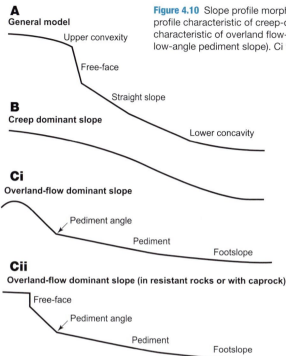

A
General model
Upper convexity
Free-face
Straight slope

B
Creep dominant slope
Lower concavity

Ci
Overland-flow dominant slope
Pediment angle
Pediment
Footslope

Cii
Overland-flow dominant slope (in resistant rocks or with caprock)
Free-face
Pediment angle
Pediment
Footslope

Figure 4.10 Slope profile morphology. **A.** Universal slope profile model. **B.** Slope profile characteristic of creep-dominated humid environments. **C.** Slope profile characteristic of overland flow-dominated semi-arid environments (note the low-angle pediment slope). Ci without caprock, Cii with caprock.

humid areas and periglacial regions, with predominantly mass movement processes, often show a marked development of the upper convexity. The bare rock free-face segment, as well as its importance in mountain regions, is best developed in periglacial and semi-arid to arid regions, where mechanical weathering is dominant over chemical weathering.

Finally, in terms of the sediment cascade, the hillslopes function as sediment source areas to fluvial systems, albeit inefficient sources, as much potential sediment released by weathering processes is stored on the hillslopes themselves (as **colluvium**). In other words, the **coupling/connectivity** of the system is far from perfect.

produced by the dominance of particular processes – creep and flowage emphasising convexity, overland flow emphasising concavity, and scree processes emphasising the free face and the (angle of rest) straight (**constant**) slope. In a very general sense, these idealised profiles (Figure 4.10) do accord with slopes generated by particular regimes of slope processess, except for slope failure and mass movement by landsliding. Because the processes are, to a large extent, climatically controlled, slope profile morphology also shows a climatic impact. Slope profiles in dry regions, with overland flow dominance, often have steep upper slopes above extensive **pediments** (*see* above, Section 4.2.1.2), the overall form marked by gentle concavity. Soil- or regolith-covered landscapes in

4.3 Fluvial systems

Fluvial systems form the core area of geomorphology. We have already seen how the drainage basin is the fundamental unit of the land surface (Section 3.2) in terms of landscape evolution at the regional scale. It is also the fundamental functional unit in terms of the sediment cascade (Figure 1.7). Flowing water in the channel network provides the primary mechanism in the long term incisional development of the landscape, but also provides the primary pathway for sediment transfer, from upland to lowland areas and from the continents to the oceans.

This is also an area of study that has advanced hugely in the last 50 years or so following the work of Luna Leopold and Gordon Wolman in applying hydraulic principles to the study of fluvial processes and landforms. They demonstrated the dependence of processes and morphology on river discharge through the concept of **hydraulic geometry**, whereby the landforms adjust by erosion and deposition to the flow conditions. The cumulatively most effective forces relate to moderate flood discharges (*see* Section 1.3), which recur from several times per year to once every few years.

4.3.1 Fluvial processes

Erosion, transport and deposition of sediment in river channels depend on the force exerted by the flowing water and the size of sediment involved. In classic work, more than 70 years ago, F. Hjulstrom demonstrated the relationship between the flow velocities required for the entrainment and the deposition of sediment in relation to sediment size (Figure 4.11). Notice how sand sizes are the first to be entrained.

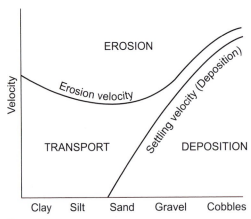

Figure 4.11 The 'Hjulstrom' curve, relating velocity to entrainment and deposition of sediment of a range of sizes.

Clays, because of cohesion between clay particles, require a higher velocity for entrainment, but once in suspension will not settle out except in what is effectively standing water. **Suspended sediment loads** travel far. Larger particles, gravel and cobbles, also require higher velocities for entrainment. They are transported as **bedload**, rolling, sliding or bouncing along the bed of the river, and there is a much narrower range of velocities before deposition takes place. Bedload moves in fits and starts sporadically through the river system.

An elaboration of these ideas is embodied in the concept of **stream power**, i.e. the power available for carrying out the work of erosion and sediment transport. The total power depends on the magnitude of flood discharge times the gradient. The power available at a point on the stream bed, the unit power, depends therefore on water depth times the gradient. Note that this is closely related to the velocity of the flowing water. William Bull developed this idea further by the concept of the 'threshold of critical power', that is, the power required to transport the sediment supplied to the stream. It is therefore sediment-dependent. If the actual power far exceeds this threshold then the stream will be erosional, transporting all the sediment supplied and cutting into the underlying bedrock. If actual and critical power are similar, there will be a broad balance between erosion and deposition, and the morphology will be adjusted to these conditions. If the critical power is much greater than the actual power – in other words if there is excess sediment supplied – then deposition will exceed erosion and the stream will **aggrade**.

These three conditions (excess power, balance, excess sediment) are expressed in the relationships between processes and morphology in fluvial systems.

4.3.2 Bedrock channels

Channels cut into bedrock are high-energy channels. Their existence is determined by the longer term evolution of the fluvial network (*see* Section 3.2), influenced by the factors that produce high effective stream power. These factors may relate either to an increase in flood power relative to sediment supply, or to an increase in gradient. The former case could follow a climatic change that radically reduces sediment supply, such as the termination of glacial or periglacial conditions. At the end of the last glaciation many rivers switched from an aggradational regime (sediment excess) to an incisional regime (sediment deficit). A change in gradient could relate to tectonic uplift or to a fall in local or regional base level. A fall in local base level could be brought about by faulting, incision of a main river to which a stream is a tributary, river capture, or by the breaching of a resistant rock band. The effects of a fall in regional base level, such as a fall in sea level, depend on the gradient of the newly exposed sea floor; incision will occur only if the gradient is sufficient.

Bedrock channels result from excess stream power, preventing the accumulation of substantial amounts of sediment. In flood conditions bedrock channels contrast with alluvial channels (*see* below) in that they normally lack any kind of flood plain. In alluvial channels, as flood discharges exceed the bankfull discharge, the water spreads onto the floodplain,

Figure 4.12 Bedrock channel morphology. **A.** Bedrock channel, Yoho Canyon, Canadian Rockies. **B.** Waterfall produced by a rock step, Aysgarth Falls, Yorkshire, England. **C.** Hanging valley waterfall, Les Ecrins, French Alps.

preventing much further increase in water depth and therefore imposing some sort of limit on flood power. In bedrock channels this is not normally the case; the channel is often constricted, especially in **canyon** settings (Figure 4.12A), and flood power continues to increase with increased flood flows. In constricted gorges and canyons, there is enormous flood power, sufficient to erode and transport huge boulders.

Bedrock channels are not equilibrium forms, in that they progressively evolve. The effects of erosion are ultimately to reduce stream power. If erosion is primarily vertical a **knickpoint** or headcut works its way upstream. As it does so it leaves behind a reduced gradient and reduced stream power. If erosion is lateral the channel widens, and depth is reduced, as is unit power. The long-term result is that unit power in bedrock channels is eventually reduced, allowing some sediment to accumulate, and the transformation of the channel from a bedrock channel to an alluvial channel.

Bedrock channel morphology reflects the nature of the underlying bedrock (massive, bedded, jointed), the gross power of the river, and the overall gradient. In the extreme case there is a step in the valley-floor profile (perhaps caused by a resistant rock bar, or by the past geomorphological history, e.g. glaciation, producing a rock step or a hanging valley situation: *see* Section 4.5). the result is a waterfall (Figure 4.12B, C). At lower gradients the profile may merely be rapids, or a relatively smooth rock-cut profile.

Where lateral erosion is important during incision, the result may be incised bedrock meanders (Figure 4.13A). Incised meanders are common in zones of moderate tectonic uplift, e.g. the Colorado Plateau, the Massif Central in France, and in the Ardennes/Rhine uplands on the borders of France, Belgium and Germany. If the meanders were cut at times of higher discharges than today, and there since has been a marked reduction of discharge to allow sedimentation on the valley floor, the result is a meandering valley within which a **misfit stream** may have developed alluvial

Figure 4.13 Incised meander morphology. **A.** Incised meanders, Tabernas, Almeria, south-east Spain. **B.** Map of a misfit stream, River Evenlode, Oxfordshire, England. Note small modern meanders within large valley meanders.

B

- Stonesfield

Meandering valley floor

- Combe

Modern (misfit) river channel

River Evenlode

- Hanborough

1 km

meanders, related to contemporary flow characteristics (*see* below, Section 4.3.3), of much smaller amplitude than the valley meanders (Figure 4.13B).

4.3.3 Alluvial channels

Alluvial channels differ from bedrock channels in that at least one bank is cut in floodplain sediments laid down by the river itself. River sediments range in size from boulders down to silt and clay. Silt and clay tend to be carried in suspension (Figure 4.14A) and are only deposited in slack water. They may be deposited by vertical accretion during overbank flows on the floodplain surface, producing laminated silts and clays. Alternatively they may be deposited in 'dead' areas within the channel. Sand tends to be transported nearer the bed, but in flood conditions can be spread onto the floodplain surface. It can form ripples or dunes or simply provide the matrix in gravelly sediments. Coarser sediments (gravels, cobbles and boulders) tend to be transported as bedload within the channel and, on deposition, form bar features. Locally, gravels deposited at the downstream end of a bar may show cross bedding, but those deposited on the bar surface show a characteristic **imbricate** fabric, with the long axes of the clasts lying across the flow direction and the major planes of the clasts dipping upstream (Figure 4.14B). Sections eroded into floodplain sediments often show basal bar sediments (sands or gravels) overlain by floodplain-surface silts or sands.

Alluvial channels undergo both erosion and deposition (Figure 4.14C). In most cases the channel morphology adjusts over a period of time through negative feedback processes (*see* Section 1.5) to the prevailing flood regime

Figure 4.14 Fluvial sediments and alluvial channels. **A.** Contrasts in suspended sediment loads between the Fraser River (behind) and the Thompson River (foreground) at Lytton, BC, Canada. The Thompson River has lost much of its sediment load by deposition in lake basins upstream of Lytton. **B.** Cobble bar alongside the South Tyne River near Haltwhistle, Northumbria, England. Note imbrication of clasts, aligned across the flow, with a dominant upstream dip; flow from left to right. **C.** Alluvial channel of the River Dane, Cheshire, England. Note that one bank (left) is erosional, while the other (right) is depositional. Note that the floodplain surface provides an effective limit to the channel, in contrast with bedrock channels such as that depicted in Figure 4.12A.

and sediment flux by erosion and/or deposition. If there is a sustained change in flood regime or sediment supply, there may be a period of net erosion or net aggradation before the channel regains a new dynamic equilibrium. In extreme cases the equilibrium may be destroyed and the channel undergoes sustained incision or sustained aggradation.

Channel size reflects the prevailing flood regime, with both width and depth increasing downstream as discharge increases, but with width tending to increase downstream faster than depth. This has implications for the cross-sectional shape of the channel (expressed by the ratio of width to depth); however, equally important is the erosional resistance of the banks. **Alluvium** deposited by rivers carrying dominantly suspended load tends to be rich in silt and clay, whereas that deposited by dominantly bedload rivers tends to be sandy. River banks composed of sandy alluvium tend to be easily eroded, resulting in wider, shallower channels in sandy material (bedload-dominated channels), whereas those composed of silt and clay have more cohesion, resulting in deeper, narrower channels (suspended-load dominated channels). Wide, shallow channels may be more effective for bedload transport, but they offer greater flow resistance, so tend to have steeper gradients. In mixed sediment load streams, banks tend to be silty, but sands and gravels are deposited as bedforms. Channel gradients reflect threshold gradients for deposition of sediment. Gradients on gravels are steeper than on sands, which in turn are steeper than on silts and clays. This is one reason for the idealised 'classic' concave longitudinal stream profile. Not only do discharges increase downstream; sediments also tend to become finer.

Figure 4.15 Secondary flows, pools and riffles. **A.** Surface expression of secondary flows, Belly River, Alberta, Canada. Note the vortices (right-hand edge of the channel, and centre left) signifying descending water and the smooth 'boils' in between, indicating upwelling water. **B.** Pool and riffle sequence. River Camel, Cornwall, England. Note the flat water surface through the pool and broken water surface over the riffles. **C.** Skew shoal development, Salinas River, California, USA. The alternate shoals develop opposite pools and are linked by riffle sections.

Channel cross-sectional shape has implications for channel pattern (the plan view of channel configuration). During flood conditions the flow does not run straight down the channel, but turns over in a series of secondary flow cells (Figure 4.15A), with descending water scouring the bed and rising water allowing deposition. In a relatively narrow single-thread channel there is an alternation between one and two such cells, but in a wide, shallow channel there are multiple turbulent cells. The cellular flow causes an alternation downchannel of scour and deposition, resulting in what is known as a **pool and riffle sequence** (Figure 4.15B), with the downstream spacing of the pools and riffles related to channel width. In a single-thread channel the pools tend to alternate from side to side of the channel (Figure 4.15C), and are associated with zones of bank erosion. The result is a **sinuous** to **meandering** channel (Figures 4.16, 4.17) (pools on the apices of the bends, shallower riffles on the crossings), whose geometry reflects the overall size and cross-sectional shape of the channel, and in turn reflects the flood regime and the sediment flux of the river. Meanders are NOT the result of obstructions; they are a natural mechanism

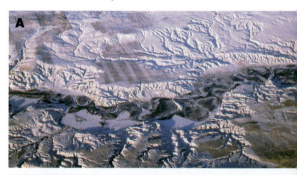

Figure 4.17 Meander morphology. **A.** Air view of the meanders of the Marias River, Montana, USA. Flow is from left to right. Note the near perfect 'double horseshoe' shape of the meanders. Note also the gullied terrain adjacent to the incised valley sides. **B.** Ground view of the meandering River Dane, Cheshire, England. The valley floor here is formed largely of a low terrace, below which are small segments of active floodplain (centre right of the photo). Note the erosional banks on the outside of the meander bends and the point bar gravel deposition on the inside of the meander bend. Note the riffle at the crossing between the two meander bends. Flow in the foreground is from left to right.

Flow direction →

Eroding bank   Pool   Meander scrolls
Point bar   Riffle

Figure 4.16 Development of a meandering pattern from a pool and riffle sequence.

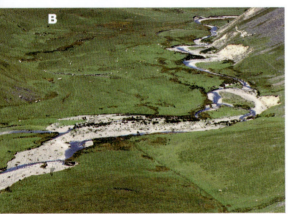

Figure 4.18 Divided channels. **A.** The gravel-bed braided channel of the Harst River, South Island, New Zealand. **B.** The wandering gravel-bed channel of Langdale Beck, Howgill Fells, Cumbria, England, after a major flood that deposited fresh gravel. **C.** Air view of an anastomosing channel, Siberia, Russia. Flow direction is towards the north, away from the camera.

whereby the river absorbs energy in the most efficient way, in extreme cases resulting in smooth 'double horseshoe' curves (Figure 4.17A). Obstructions tend to distort meanders.

In wider, shallower, bedload-dominated channels the flow patterns are more complex, resulting in multiple patterns of scour and shoaling, forming sand bars or gravel bars interlaced by a **braided channel** network

(Figure 4.18A). At low flows mobile sand or gravel bars become emergent, and may later be colonised by vegetation, becoming more stable islands. There are two mechanisms characteristic of braiding; the first involving channel widening and the deposition of mid-channel bars (primary braiding); the second involving spillage of the floodflows onto the floodplain and the re-occupation and scour of older abandoned channels within the flood-plain (secondary braiding). In most braided rivers both mechanisms operate. Braiding is a response to large volumes of coarse sediment input, but is not necessarily indicative of aggradation. Given that gradients in wide, shallow channels are high, braiding is the most effective channel form for transporting large volumes of coarse bedload sediment.

There are several departures from these two idealised channel patterns. In single-thread channels, if gradients are too low or the banks are too resistant to erosion to allow meander development, then a narrow, mildly sinuous non-meandering channel is the result.

A transitional channel type between meanders and braids, common in upland bedload-dom-inated systems, is the wandering gravel-bed river (Figure 4.18B). The channel is wider and shallower than a typical meandering channel and tends to wander in a pseudo-meandering fashion. The bars tend to be larger than the point bars on the inside of meander bends, but are dominantly side bars rather than the mid-channel bars typical of true braiding.

There is a second type of multiple channel, which charac-terises low gradient, mud-dom-inated systems. These operate more as several discrete channels rather than as one braided channel. The individual channels may well meander, but they are relatively stable and migration rates are generally low. They tend to be aggrading channels, often

bounded by levees, beyond which the flood-plain is often in the form of backswamps and at a level lower than the aggrading channel floor. Channel change is generally brought about by spillage during flood conditions (**avulsion**), breaching the levees and estab-lishing a new channel branch through the backswamp floodplain. These are known as **anastomosing** channels (Figure 4.18C). They are relatively low energy channels and should not be confused with the much higher energy braided channels. Similar channels are often found in estuaries.

In some classic work in the 1950s Leopold and Wolman identified threshold conditions

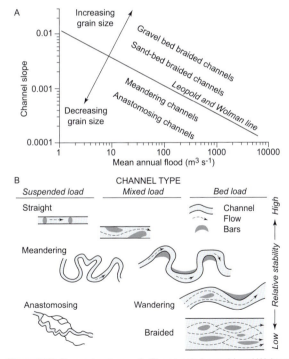

Figure 4.19 Channel patterns. **A.** The classic Leopold and Wolman diagram differentiating meandering and braided channels on the basis of floodflows and channel gradients, with plotting position of anastomosing channels added. **B.** Schematic representation of channel pattern types.

between meandering and braided channel regimes based on flood discharge and gradient. Later research has incorporated other channel types into this concept, together with the role of sediment (*see* Figure 4.19A, B).

4.3.4 *Alluvial depositional forms*

4.3.4.1 *Floodplains*

The **floodplain** is an integral part of an alluvial river system. It is composed of fluvial sediment stored on the valley floor. There are various types of floodplain, related to the channel style. In meandering and braided systems, where channels are subject to lateral migration, the floodplain is continually being added to by point-bar deposition and island stabilisation, and removed by bar migration and bank erosion. On most such floodplains the surface may comprise segments of different ages (Figure 4.20). The floodplain sediments also store a sedimentary record of flood history and channel behaviour. In dominantly accretionary floodplains there may be long periods of aggradation and perhaps occasional removal by rare catastrophic floods.

The floodplain also acts as a physical limit to the channel, when it is temporarily inundated by overbank flood flows. This tends to reduce

Figure 4.20 Floodplains. **A.** Floodplain of the meandering River Dee, near Llangollen, North Wales. Note the multiple segments on the floodplain surface. **B.** Floodplain of the Thompson River, British Columbia, Canada. Note the flood channel across the bend. Note also how different vegetation picks out age and sediment variations within the floodplain. Flow is from left to right.

the severity of flood peaks downstream, but it also limits the available stream power at a cross section. From the point of view of river and land management, floodplains have to be able to flood. The floodplain is not a sensible place to build!

4.3.4.2 Alluvial fans

Alluvial fans are rather different depositional landforms from floodplains. By definition they cannot be equilibrium landforms, in that they are the result of net sediment accumulation. They represent much longer-term sediment storage. Alluvial fans occur

where sediment-laden streams enter a zone where sediment transport capacity is much reduced, so that sediment, especially the coarse fraction, is deposited forming a conical depositional feature whose slopes radiate away from the apex. There are two important topographic situations where alluvial fans are common (Figure 4.21). One is at a linear mountain front, which may be fault-bounded or simply a mountain-pediment junction zone. The other is at a tributary junction where a tributary valley joins a major valley.

Alluvial fans are common in dryland mountain regions, but occur in all climatic mountain or upland zones. They range enormously in size, from features up to tens of kilometres in length (megafans), such as those at the front of the Himalayas, to small cones a few tens of metres in length. Similarly, depositional processes vary, ranging from debris flows to fluvial deposition by sheetflows or by braided, meandering or anastomosing channels. Fan surface morphology adjusts to the prevailing flood and sediment regime. Debris flows are deposited at a relatively steep gradient, unconfined sheetflows deposit sediments at intermediate gradients, and channelised streamflows at lower gradients. If the flood and sediment regime changes over

Figure 4.21 Alluvial fans. A. Mountain-front fans along the uplifted and tilted Panamint Range on the west side of Death Valley, California. Note the complex of fan segments; also the playa lake on the floor of Death Valley below. B. Mountain-front fan, Grotto Canyon Death Valley, California, USA. Note the multiple age fan surface segments. C. Tributary-junction fan complex, Musandam Mountains, Oman. Note the steep debris cones also feeding sediment to the tributary-junction fan, and the tributary-junction fan feeding sediment into the main wadi channel. Flow in the main wadi is from left to right [For other illustrations of alluvial fans *see* Figures 3.1B and 5.2B].

A

B

**FANS WITH LITTLE OR NO CURRENT EROSION
OR DEPOSITION**

PASSIVE / INACTIVE

FANS DOMINATED BY DEPOSITION

DEBRIS-FLOW DEPOSITION

SHEET FLOOD OR
FLUVIAL DEPOSITION

COMPOSITE
DEPOSITION

	Sediment supply	
	Low ⟷ High	

Water supply	High	DISSECTION / PROGRADATION
		DEPOSITION
		Fluvial and sheetflood (composite)
	Low	PASSIVE/ INACTIVE Debris-flow

Figure 4.22 Style of alluvial fans (**A**) in relation to flood power and sediment supply (**B**).

FANS EXHIBITING COMBINATIONS OF EROSION AND DEPOSITION

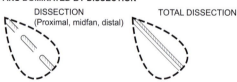

PROGRADATION

(On steep cones,
may involve debris flows)

FANS DOMINATED BY DISSECTION

DISSECTION
(Proximal, midfan, distal)

TOTAL DISSECTION

time the fan morphology will adjust through erosion and deposition, modifying the fan gradient.

Several fan styles can be identified (Figure 4.22A). On an aggrading fan, sediment is deposited from the apex down. Another common form is the **prograding** fan, where in proximal zones the **axial channel** (that is,

the main fan channel, carrying water and sediment from the feeder catchment, and normally running down the axis or centre of the fan) is incised below the fan surface in a fanhead trench. This converges with the fan surface in midfan at an intersection point, below which deposition is dominant, often extending the distal limits of the fan. Finally the fan may be undergoing dissection, not only within the fanhead trench, but also in midfan below the intersection point, or, when associated with a local or regional base-level change, in the distal zones.

Within the context of the topographic setting, fan style adjusts to the prevailing flood and sediment regime (Figure 4.22B). A change in this regime, either an abrupt or a progressive one, or a change in base level can cause a change in fan style. In this way the sediments and morphology of fans preserve a sensitive record of tectonic or base-level changes in their settings, or of environmental change within their source areas, such as alterations in climate or land cover.

4.3.5 Fluvial change

The behaviour of alluvial fans, described above, is just one example of how relatively small parts of the fluvial system respond to environmental change. On a much larger scale the whole fluvial system is susceptible to change. Within the context of the overall valley form and gradient inherited from the geomorphological past, alluvial channel systems tend to adjust their form to the prevailing flood and sediment regime. Fluctuations in flood regime or sediment supply may cause changes in channel geometry and pattern, but much larger 'threshold' changes may be too great to be accommodated in this way and may require a radical change in gradient involving sustained aggradation or incision.

The causes of such threshold changes include gradient changes and changes to the flood or sediment regime. Gradient change may be brought about by tectonics, by post-glacial isostatic uplift, by base-level change or by human activity such as channel straightening (see Chapter 6). Most significant are changes in flood regime or sediment supply, which can be both climatically or human-induced (see Chapter 6).

Responses to gradient change tend to be localised. An increase in gradient may trigger incision, whereby a knickpoint or headcut forms in the channel floor and propagates (albeit relatively slowly) upstream. A decrease in gradient may trigger local aggradation, but significant propagation upstream is negligible. The river upstream 'has no idea' what happens downstream!

Responses to changes in flood hydrology or sediment supply tend to affect a greater part of the system more rapidly. Changes in flood hydrology are normally propagated downstream, including reductions in flood peaks brought about by river regulation. Increases in sediment supply are propagated downstream; immediately in the case of suspended sediment; more slowly in the case of coarse sediment that may cause a 'sediment wave' to move through the system.

Channel responses to a radical reduction of effective stream power or to sediment overload are sustained aggradation, with burial of the old valley floor and eventual deposition of a stack of sediments of much greater thickness than the normal maximum depth of the river. The opposite case, a radical increase in effective stream power, is sustained incision. The old valley floor is abandoned as a terrace somewhat above the level of the new incising channel (Figure 4.23; see also Figure 3.11B). Both aggradation and incision result in a change in gradient. Aggradation of a sediment wedge increases the valley-floor gradient. Incision may create a headcut that migrates upstream, leaving behind a channel of a

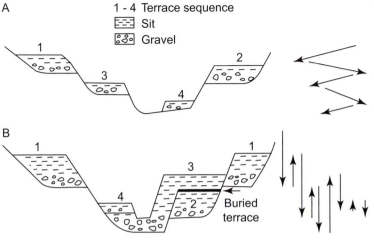

A

1 - 4 Terrace sequence

Sit

Gravel

B

Buried terrace

Figure 4.23 Schematic diagram illustrating the development of river terraces. **A.** Terraces produced by lateral migration during progressive incision. **B.** Terraces produced by alternating incision and aggradation.

lower gradient than the now dissected valley floor. Again, responses to both aggradation and incision are examples of the stabilising effects of negative feedbacks. These changes of gradient tend to enhance the development of a new equilibrium form in adjustment to the new regime.

Sediment preservation and the relationships between morphology and sediments in **river terrace** sequences provide important evidence to support the interpretation of the response of the geomorphic system to past environmental change. The relationships between the erosional base of a river terrace, the thickness of the deposits and the elevation of the river terrace surface (Figure 4.23) can provide evidence of past sequences of erosion and deposition. The sediments themselves can tell us about palaeo-river morphology, and the degree of soil development on the terrace surface may tell us something about the age of the landform (*see* Section 5.2). We have already introduced the valley filling and channel entrenchment of arroyo systems in

the American West and seen how they may be viewed as terrace sequences produced by aggradation and incision partly as responses to environmental change (*see* Section 4.2.1.2).

Terrace sequences are widespread elsewhere. In many young mountain regions terrace sequences carry a tectonically-induced signal. In the temperate latitudes of the major European river systems the terrace sequences reflect the Pleistocene glacial/interglacial sequence, within the context of overall incision of the river systems during the Pleistocene. Aggradation tended to occur under periglacial conditions during the glacials, dominantly by braided river systems. Sediment supply was much reduced during the interglacials, resulting in incision, followed by the development of a new floodplain, generally by meandering rivers.

Over the shorter timescale of the period since the melting of the ice from the Last Glacial Maximum within upland northern Britain, there have been significant phases of river terrace development. The periglacial conditions of the late Pleistocene saw the aggradation of the valley floors throughout much of upland northern Britain. The sediment starvation, resulting from the cessation of periglacial processes and the early **Holocene** vegetation colonisation of the hillslopes, saw the incision of the river systems into the late Pleistocene valley floors to form river terraces along most of the major rivers of the region. Later on, during the Holocene, smaller, younger terraces were formed in response to climatic variations and human-induced changes in sediment supply (*see* Chapters 5, 6). Post-glacial isostatic rebound has often been cited as a possible cause of the late Pleistocene to Early Holocene incision and terrace development. It is true that this might have steepened gradients in the Scottish Highlands, but would have been less effective further south. Furthermore, terraces are evident on river systems that flow towards the zone of isostatic uplift as well as on those that flow away.

4.4 Aeolian systems

Perhaps the least widespread of the major sediment systems, and that with only limited geomorphic impact, is the aeolian system – sediment entrainment, transport and deposition by the wind.

4.4.1 Aeolian processes

Sediment entrainment by the wind (with the exception of the entrainment of volcanic dust supplied directly by volcanic activity) takes place under strong wind conditions blowing across loose, dry surfaces. Entrained sediment sizes range from dust and silt to sand. Once entrained, dust and silt may be carried high in the atmosphere over long distances, but sand is generally carried much nearer the ground and over much shorter distances. Only sand

has the capability of effecting erosion by 'sand-blasting' rock features. Nevertheless, this does produce wind-etched forms (**yardangs**) in some desert areas. Otherwise, in aeolian geomorphology we are dealing with the entrainment, transport and deposition of dust, silt and sand.

There are several dust sources. These include soils and palaeolake surfaces in deserts, especially in the Sahara and central Asian deserts. Saline and carbonate-rich dust from palaeolake surfaces, when redeposited, forms an important constituent of desert soils. Agricultural land, especially in drylands, is another important source of atmospheric dust. In extreme cases, such as during the 'dirty thirties' in the American Great Plains, such dust may be transported by dust storms long distances from source and deposited as a blanket over large areas. Another important source of dust is from volcanic eruptions. Deposition from dust clouds has limited impact as a geomorphological phenomenon, but the deposition of volcanic ash may have particular significance in dating landform and sediment sequences. Each eruption tends to produce ash with a diagnostic mineral assemblage. For example, the ash from the Mount Mazama catastrophic eruption in 6600 BP, which created Crater Lake caldera in Oregon, was spread over large areas of the American West and can be recognised in sediment sequences by its particular mineralogy.

Perhaps more important in terms of scale are the Pleistocene loess sheets (*see* Section 2.2.3; Figure 2.3), deposited over large areas of the American Midwest, the north European plain and central Asia. The source of the silt was glacial meltout from the continental ice sheets over Eurasia and North America. The result was a depositional blanket up to tens of metres thick, but in the Loess Plateau of China its thickness exceeds 200 m. The Chinese loess has preserved a record of climate conditions extending back to the early Pleistocene, in the form of superimposed loess sheets interstratified with palaeosols. Loess deposition has left fertile but, when dissected, easily erodible soils. The Chinese loess plateau is one of the most spectacular gullied areas in the world, and the Yellow River, fed by this erosion, is one of the most sediment-laden rivers in the world.

Geomorphically, the most important sediment transported by aeolian processes is sand. Sand transport takes two forms, both dependent on a windspeed threshold, which is lower for dry than for moist grains; surface creep, where the sand grains move by rolling and sliding along the surface; and **saltation**, where the grains are supported within the airflow at heights of up to about 1 m from the surface, and travelling distances of up to several tens of metres.

4.4.2 Aeolian depositional morphology

On deposition, aeolian sand forms a variety of landforms from sand sheets to rippled sand, and most importantly, to a variety of dune forms. Dunes are important in two zones, on depositional coasts (*see* below, Section 4.6) and in deserts.

The sand source for coastal dunes is the beach exposed at low tide, so dune coasts preferentially occur where there is a high tidal range on gently shelving shorelines. Coastal dunes tend to have a higher moisture content and are less mobile than desert dunes; moreover, they are more subject to colonisation by vegetation (especially by marram grass). They are subject to 'blowout' which exposes fresh sand

to aeolian activity. Characteristic dune forms are transverse dunes, parallel with the coast, and parabolic dunes (crescentic features whose arms point into the wind) (Figure 4.24). At times of falling sea level, when unconsolidated sediments may be exposed, dunes may form on coasts where dune systems are otherwise uncommon. Around the western Mediterranean, where today dunes are uncommon, there are cemented fossil dunes (composed of **aeolianite**) dating back to falling sea levels during the Pleistocene.

We think of dunes as the typical desert landscape (Figure 4.25), but only about 25% of the world's deserts are occupied by dunes. Desert dunes depend on the availability of sand. Much of this comes from terminal fluvial deposits or from fluvial deposits related to river systems that were functional during times of less aridity (e.g. during the late Pleistocene in the Sahara). Elsewhere, the sea floor, exposed at times of low sea levels, may have been an important sand source (for instance, during much of the Pleistocene the Arabian Gulf was dry, providing some of the sand for the dune systems of the UAE.).

Dune morphology reflects sand supply and wind direction (Figure 4.24). With a single dominant wind direction there is a family of dune forms ranging from simple **barchans** to barchanoid ridges to transverse dunes, related to increasing volumes of sand available. All barchan forms have a smooth upwind face and a steep downwind avalanche face that develops at the angle of rest of the sand. The 'wings' of barchans point downwind. The enormous 'megabarchans' of some parts of the Sahara, the Arabian and Taklimakan (China) Deserts probably formed in the late Pleistocene, under conditions of a high rate of sand

Parabolic (crescentic) dune

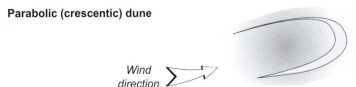

Wind direction

Figure 4.24 Dune types in relation to wind directions and sand supply.

Barchan dune

Linear dune

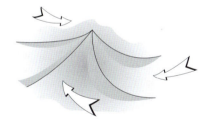

Figure 4.25 (below) Desert dunes. **A.** Part of the large sand sea ('The Empty Quarter') near the border of the UAE and Saudi Arabia. Barchanoid dunes – main wind direction is from right to left. Note the steep avalanche (slip) face of the dune on the right of the photo; note also the rippled sand in the foreground. **B.** Close up view of climbing barchanoid dunes near Liwa, UAE. Wind direction is from right to left. Note the left-facing slipfaces of the dunes.

Star dune

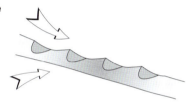

supply and a dominant single wind direction. Between the megabarchans of the southern UAE are Pleistocene '**sabkha**' deposits, remnants of ephemeral saline lakes. Linear dunes, common in Australia and southern Africa, result from the interplay of seasonal variations of wind direction oblique to the plan view of the dune (Figure 4.24). More complex forms, such as 'star dunes', result from the interplay of multiple wind directions.

In addition to today's active desert dune-fields, there are many areas of Pleistocene dune activity that have now stabilised. Stabilisation of dunes takes place as sand supply diminishes and an algal crust forms on the dune surface, followed by invasion by higher plants. One such area is the Sand Hills area of Nebraska, today a long way from arid zones. The source of sand was probably fluvial sediment from glacial rivers. Elsewhere, for example in the Sahel and in Botswana in Africa, and in several areas in Australia around the active dunefields, there are now vegetated and stabilised dunes that suggest a much greater extent of aridity or of greater sand availability at times during the Pleistocene.

4.5 Glacial systems

Glacial ice has produced the most spectacular mountain scenery in the world. Although the present glacial extent is limited, during the Pleistocene large areas of the northern continents were covered by glacial ice, which effectively created the landscapes of large areas of North America and northern Eurasia (*see* also Figure 2.3). Glaciers and ice sheets range in scale (Figure 4.26) from small niche and cirque glaciers in mountains, to valley glaciers, plateau icefields, and to continental-scale ice sheets, the largest of which during the Pleistocene was the Laurentian ice sheet, which covered most of Canada and a significant portion of the northern USA.

4.5.1 Glacial and fluvio-glacial processes

Glaciers form where, over a sustained period, the annual snow accumulation exceeds the summer snowmelt. The snow accumulates, is compacted into '**firn**' (white ice containing air), then after a number of years into true glacial ice, with few or no air bubbles. Ice has

Figure 4.26 Glaciers. **A.** A small niche glacier. Crowfoot Glacier, Canadian Rockies, Alberta. Note the ridge in the bottom right of the photo. This is the Little Ice Age (LIA) moraine marking the late Holocene maximum extent of the glacier, reached about 200 years ago. **B.** A valley glacier, Nisqually Glacier, Mount Ranier, Washington, USA. Note the snow-covered ice above the equilibrium line (mid-photo) and the dirty ice below. The pronounced ridges on either side of the glacier mark the LIA maximum extent of the glacier. **C.** Detailed view of the ablation zone of the Rhone Glacier, Switzerland. Note the crevasse patterns. **D.** Detailed view of the snout of the Athabaska Glacier, Canadian Rockies, Alberta.

two styles of deformation: under atmospheric pressure it behaves as a rigid elastic solid, and fractures under stress. Under greater pressures, i.e. at depth within a glacier, ice behaves as a quasi-plastic material and deforms by slow flowage. Furthermore, under pressure ice may reach its melting point, so that at their bases many glaciers are near the pressure melting point (so-called 'temperate' glaciers, or warm-based glaciers), allowing the basal layers to melt and refreeze. This is important for glacial mobility and for the erosion of rock

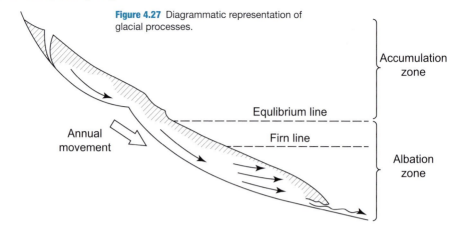

Figure 4.27 Diagrammatic representation of glacial processes.

from the base of the ice. On the other hand, the temperature at the base of cold-based glaciers is way below the pressure melting point. Such glaciers are frozen to the substrate below, are much less mobile, and less effective in erosion.

As the ice builds up in the accumulation zone (Figure 4.27) it begins to flow down any topographic gradient. In a continental ice cap the ice will build up to form an ice dome, completely burying the underlying topography. In mountain glaciers the accumulation zone may be high altitude hollows, as in the case of many mountain **cirques**, or on a mountain plateau, such as that of the Columbia icefields in the Canadian Rockies, or of the Norwegian icefields. The ice flows down into the zone of net **ablation**, in mountain glaciers as outlet or valley glaciers, to where annual melt exceeds annual accumulation. In an ice sheet, flow will be away from the topographic high of the dome of ice. The position of the ice front will depend on the balance between the rate of glacier flow, itself dependent on gradient and the mass of ice, and on the melt rate, especially at the margin of the ice sheet or the snout zone of a valley glacier. Climatic fluctuations,

changes in snowfall, and changes in temperature will influence whether the glacier front advances or retreats. There will be a time lag between changes in **mass balance** (the net balance between addition by snowfall and loss by melting) and advance or retreat of the glacier front, and because of different topographic settings the behaviour of neighbouring glaciers may not be synchronous.

As the ice moves, the surface layers deform by fracture, forming crevasses in zones of extension and being thrust upwards in zones of compression. As the ice moves, sediment is entrained in two main ways; by direct erosion of the underlying rock below and at the sides of the glacier, and in mountain glaciers by the addition of sediment onto the surface of the glacier from rockfalls and other mass movement processes (**supraglacial moraine**). The underlying material is entrained partly by 'plucking' rocks frozen into the bed, partly by the hydraulic action of subglacial meltwater, and partly by the abrasive action of the rocks already embedded in the ice. The rocks within a matrix of silt derived from abrasion form the **englacial moraine** within the lower part and at the margins of the glacier. Meltwater is also important, both subglacially in warm-based glaciers, and from the surface of the ice in the ablation zone. The surface meltwater may form streams on the ice surface, but disappear into the glacier through crevasses (**moulins**), to emerge at the glacier snout as a meltwater river.

There are several processes involved in glacial deposition. On deposition most morainic material (**glacial till**, also known as **boulder clay**) is very poorly sorted and may contain all grain sizes from boulders to silt and clay. Englacial moraine deposited beneath an active glacier or ice sheet is compacted and dense (**lodgement till**), in contrast with that deposited at the glacial margins during glacial melt (**meltout till**). Where flowing water is involved, either englacially or **proglacially**, the sediment may be better sorted and comprise sands and gravels. A distinctive environment occurs if the ice is cut off from source and no longer mobile (**dead ice**), in which case the deposits are simply dumped as the ice melts and collapses.

Meltwater itself is important, both englacially and proglacially. Glaciers can temporarily dam lakes. The well-known parallel roads of Glen Roy in Scotland, first described by Agassiz, are preserved shorelines of a series of late Pleistocene ice-dammed lakes. Ice-dammed lakes may drain either into the ice or externally, and can cause catastrophic outburst floods. Enormous proglacial lakes formed around the margins of the Laurentian Ice sheet during late Pleistocene deglaciation (including Lake Agassiz), and were the forerunners of the modern Great Lakes and the large Canadian lakes further north. The most catastrophic outburst flood, indeed the greatest ever flood, was the Lake Missoula outburst flood that occurred in the Pacific Northwest of the USA, and created the channelled scablands of Washington State, huge fluvially cut features. More generally, though, meltwater rivers issuing directly from a glacier snout are highly charged with silt and transport coarser sediment into the foreland zone or proglacial valley, where they form unstable braided rivers that run through the zone of outwash sediments known as a **sandur**.

4.5.2 Glacial and fluvio-glacial erosional forms

Landforms produced by glaciation include both erosional and depositional forms. Erosional forms produced by ice sheets include extensive glacially scoured bedrock surfaces. The core areas of both the Pleistocene Laurentian and Scandinavian ice caps coincided with the outcrop of Precambrian shield rocks (*see* Chapters 2 and 3), composed of erosionally resistant high-grade metamorphic rocks. The resulting landscape left after glaciation is one of huge areas of smoothed and scoured bedrock (Figure 4.28A), sometimes in the form of ridges with a 'plucked' distal end (a **roche moutonnée**) and at a detailed scale may show grooves cut into the rock (Figure 4.28B). Between the ridges are glacially scoured hollows that coincide with faults, rock joints and other zones of erosional weakness. These have since become lakes and peat-filled basins. The landscape as a whole has a totally deranged drainage system. In Britain there is a similar landscape of glacial scour on the Lewisian, Precambrian metamorphic rocks of the far northwest of Scotland (*see* also Figure 3.9C).

Meltwater associated with ice caps cuts erosional channels, both subglacially and

Figure 4.28 Glacial erosional landforms. **A.** Large-scale glacial scour of rock surfaces by ice sheet glaciation. Aerial view over Labrador, Canada. The direction of ice movement was obliquely from right to left towards the camera. **B.** Glacially scoured rock surface, Yosemite, California, USA. Ice flow was from right to left. **C.** Aerial view of glacially erosional mountain landforms, Swiss Alps, south of Zurich. Note back-to-back cirques, and arête ridges. **D.** Cirques above the Bow Valley, Canadian Rockies, Alberta. **E.** Glacial trough and hanging tributary valleys, Yosemite Valley, California, USA.

ice-marginally. On deglaciation some of these channels may form the primary valleys of the post-glacial fluvial network; others may be totally abandoned, but be preserved as anomalous features. The erosion of subglacial fluvioglacial channels is also one of the mechanisms whereby glacial diversions of drainage systems can be produced (*see* Section 3.2.1).

Erosional forms produced by mountain glaciers are much more distinctive (Figures 4.28C, D, E). Any thickness of ice will tend to scour vertically to form hollows and basins, such as high altitude cirques and overdeepened valleys. Back-to-back cirques may cut back to form an **arête**, a knife-edged ridge separating the two headwalls. Multiple cirques may cut back to form a pyramidal peak; the Matterhorn on the Swiss–Italian border is the classic example. Valley glaciers transform pre-glacial V-shaped fluvial valleys into U-shaped glacial valleys. Deepening of the main valley floors may leave the tributary valleys as **hanging valleys**. Overdeepening of the valley floors may, on deglaciation, result in finger lakes, or in coastal areas, fjords.

4.5.3 Glacial and fluvio-glacial depositional forms

Glacial depositional forms (Figure 4.29) include both glacial and fluvio-glacial forms, produced subglacially, at the ice margins, and proglacially. Subglacially, lodgement till may be smoothed into streamlined whaleback forms (**drumlins**). These may be purely 'drift' features or may be rock-cored. Around the margins of active ice, **morainic** ridges may be produced, sometimes simply composed of meltout till, sometimes as push moraines thrust forwards by an advancing glacier. **Dead ice** produces its own suite of depositional forms, including: hummocky moraine formed as the ice melts and collapses; piles of loose material from crevasse fillings (**kames**); hollows left where detached blocks of ice melted (**kettle holes**); terrace-like forms

Figure 4.29 Glacial depositional landforms. **A.** Fresh moraine at the margins of the Athabaska Glacier, Canadian Rockies, Alberta. Note unsorted nature of the sediment. **B.** Section in late Pleistocene boulder clay (till), exposed in the South Tyne valley, near Alston, Cumbria, England. **C.** Late Pleistocene drumlins, the smooth low hills in the centre of the photo, Ribblehead, Yorkshire, England. **D.** Dead ice topography, Drumochter Pass, Scottish Highlands. Note the hummocky morainic terrain in the centre of the photograph. This was deposited by a tongue of dead ice during the Loch Lomond glacial phase at the very end of the last glaciation. The whole area was covered by ice during the Last Glacial Maximum, but the smoother hillslopes above the hummocky moraine of the Loch Lomond ice were modified by periglacial processes during the late Pleistocene.

that accumulated at the margins of a down-wasting glacier of dead ice (**kame terraces**); and sinuous linear ridges of sand and gravel (**eskers**) marking the courses of subglacial streams. As described above, meltwater rivers create spreads of outwash gravel (**sandur systems**).

These forms are present under, in and around modern glaciers, but they also have significance as landforms in areas glaciated by ice sheets and mountain glaciers during the Pleistocene.

4.6 Coastal systems

Coasts offer some of the most dramatic and dynamic landscapes. The coastal environment includes high energy zones, where processes are more or less continuously active, but also low energy zones. In coastal environments terrestrial and marine processes interact. Most coastal sediment is ultimately derived from terrestrial sources, river systems or cliff erosion, and is reworked by marine activity. Coastal landscapes are young; world sea levels reached their present level only about 6000 years ago.

Coastal geomorphology encompasses not only marine coasts but also coasts on inland bodies of water, where the processes are similar, but more limited than those on sea coasts.

4.6.1 Coastal processes

Marine coastal geomorphology is influenced by the underlying geology and by the regional geomorphic history. There are marked differences between tectonically active and passive coasts, partly expressed by the differences between the generally straight tectonically-active coastlines around the North Pacific Ocean and the much more irregular, tectonically stable coastlines around the North Atlantic. However, some of these differences and some of the exceptions relate as much to glacial as to tectonic history.

The dynamic marine coastal processes are driven by three main phenomena: waves, tides, and currents. Waves are generated by wind-induced stresses on the open sea, which cause an in-situ rotation of the water (Figure 4.30). This in turn generates a wave form that propagates with the wind direction and amplifies with distance (**fetch**) and wind strength. For this reason, wave amplitude and therefore wave power is potentially much greater on

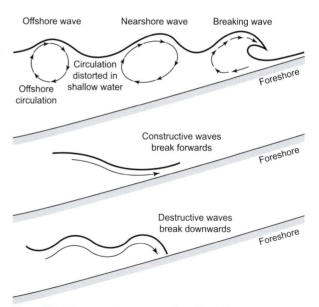

Figure 4.30 Diagrammatic representation of breaking waves contrasting constructive and destructive waves.

open ocean coasts than on constricted seas. Consider the location of the world's greatest surfing beaches!

As the waves approach the shore, the water rotation is distorted by the drag exerted by the shallowing water, so that the upper part of the wave is propelled shorewards. In other words the wave begins to break. At this point another consideration becomes important; that of wavelength. Gentle offshore profiles tend to increase the wavelength of the breaking waves, whereas steep profiles tend to maintain a relatively short wavelength. The wavelength influences the style of breaking as the wave reaches the shore. Waves with a long wavelength, especially when breaking on a gently sloping shore, tend to break sub-horizontally up the shore and are referred to as constructive waves, whereas those with a short wavelength, especially when breaking onto a steep shore, tend to break downwards and are referred to as destructive waves (Figure 4.31).

There are a number of complications. On an irregular deepwater cliffed shore, waves can be reflected by the shore and interfere with one another, either damping or reinforcing wave amplitude. Wave approach to the shore is usually oblique, which has implications for sediment movement (*see* below), but may be altered by wave refraction from an irregular offshore bed profile.

Tidal range also affects shoreline processes. Tides are generated primarily by the gravitational attraction of the Moon on the world's oceans as the Earth spins on its axis, resulting in two tidal cycles per day. Coastal configuration modifies the oceanic tides; on some coasts increasing, and on some coasts reducing the tidal range. The picture is further complicated by the monthly lunar cycle, so that twice a

Figure 4.31 Waves. **A.** Destructive waves (short wavelength, breaking downwards) breaking onto a shingle beach, Sidmouth, Devon, England. **B.** Constructive waves (long wavelength, breaking forwards) breaking across a wave-cut platform, near Hartland, Devon, England.

month, when the gravitational attraction of the Moon and the Sun reinforce each other, the tidal range is greater (spring tides coinciding with the new and full moon) as opposed to the intervening periods when the two effects reduce the tidal range (neap tides). Tidal range is further modified annually, with equinoctial tidal range (March, September) being greater than at solstice times (December, June).

The effects of tidal range are to modify the vertical extent of shoreline processes. For example, in the Mediterranean Sea, with its almost negligible tidal range, beaches are narrow and coastal erosion is restricted to a very narrow vertical zone, whereas on the British coasts of the Irish Sea, beaches are wide and coastal erosion is effective over a wider vertical range. The highest tidal range of all is in the Bay of Fundy in eastern Canada. Tidal range also affects estuarine and delta processes, so that marine effects are propagated further inland on coasts with a high tidal range.

Ocean currents have little direct effect on coastal processes, except on water temperature, and therefore on biological activity. However, local currents, generated by river inflows or by tidal flows, can influence sediment movement.

Two other effects should be mentioned. Storm surges generated by weather conditions may temporarily raise the tidal zone, with implications for coastal flooding. In tectonically active regions earthquakes may trigger tsunamis, which can have a devastating effect on coastal environments and may be transmitted across wide oceans to affect the opposite shores.

4.6.2 *Erosional coasts*

On steep, high energy coasts wave attack is concentrated at the base of the seaward slope, creating first a notch at the slope base, then a low cliff as the notch is enlarged and the overhanging rock falls (Figure 4.32). Continued wave attack at the base of the cliff causes the cliff to recede, producing a distinctive morphology. At the base is a **wave-cut platform** (Figure 4.33A), a seaward sloping surface cut into the rock as the cliff recedes. This culminates at the base of the cliff in a notch (Figure 4.33B) and active cliff, above which there may be a seaward slope on which subaerial processes, gullying, and mass movements including landslips may operate (*see* Section 4.4; Figure 4.9B). The detailed form varies with the original relief into which the cliff is cut, the geological structure and the rock resistance to erosion. Isolated zones of stronger rock may produce sea stacks (Figure 4.33C); weaker zones, faults and joints, may produce caves and arches.

As the cliff recedes, producing a wider wave-cut platform, some of the wave energy will be dissipated on the platform, reducing

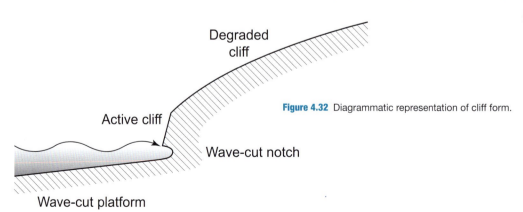

Degraded cliff

Active cliff

Wave-cut notch

Wave-cut platform

Figure 4.32 Diagrammatic representation of cliff form.

Figure 4.33 Cliff forms. **A.** Irregular cliffed coastline with headlands and sea stacks, Lizard Point, Cornwall, England. **B.** 'Wave-cut' notches in limestone cliffs, Calpe, Spain. These notches probably owe their form more to the phytokarstic effects of algae that colonise the waterline, than to wave power alone. **C.** Exposed wave-cut platform, cut across steeply dipping bedrock, near Hartland, Devon, England.

that expended at the base of the cliff, allowing sediment to accumulate in the form of a beach below the cliff. This is a very common situation where wave attack at the base of the cliff occurs only during storm conditions under high tide conditions.

The effects of cliff erosion are that headlands suffer greater erosion then embayments, ultimately to reduce the complexity of the plan view of the coastline, and ultimately to replace cliff-only coastal segments by cliff-and-beach segments.

4.6.3 Depositional coasts

Depositional coasts range from the cliff-and-beach coasts, described above, both 'cape and bays' coasts and straight coasts, to coasts dominated by depositional features. In all cases the depositional shoreline is fringed by a beach of loose sediment. The sediment sources are ultimately largely terrestrial, cliff erosion or river sediment. Even sediment driven onshore as post-glacial sea levels rose may have been largely reworked river sediments, deposited during lower sea levels on the then exposed sea floor. Beach sediment sizes are dominantly sand and shingle, silts and clays only being deposited in low energy highly sheltered coastal environments, such as in estuaries and coastal lagoons. Silt and clay are only deposited on open coasts on very low energy coasts, and then primarily in association with vegetation in salt marshes or, in the tropics, mangrove (*see* below). Beach sediments are generally mature, well sorted sands and shingle. Beach pebbles are generally well rounded and often show a concentration of the more resistant lithologies, quartz and flint being common, the weaker lithologies having been broken down into the sand fraction.

Beach sediments are constantly being reworked, being washed up the beach on the incoming tide under constructive wave action and along the beach by **longshore drift** (Figure 4.34). This is the process related to the usual oblique approach of waves to the

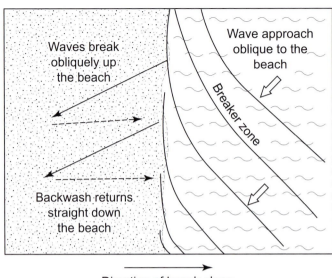

Waves break obliquely up the beach

Wave approach oblique to the beach

Breaker zone

Backwash returns straight down the beach

Direction of beach slope

beach, which tends to wash sediment both up and along the beach. The return 'backwash' returns the sediment down the beach only for it to be taken obliquely along the beach with the next wave. On most beaches there is clearly a dominant direction of drift related to the dominant wind direction and the usual direction of wave approach. We often see evidence of this where groynes have been constructed for beach protection. Sand is usually piled up on one side of the groyne. Even though the origin of many sand and shingle structures may have been the onshore movement of sediment during the post-glacial sea-level rise, their present form has been modified by longshore drift, whereby spits and bars extend sub-parallel with the coast, often protecting quieter, lower energy environments behind them (Figure 4.35A). Spits are especially common at river mouths, sometimes totally enclosing a lagoon, but shingle features also occur on more open coasts, forming complex depositional headlands.

There is often a vertical sorting of beach sediment, an upper shingle beach above flatter intertidal sand. On exposed coasts, shingle storm beaches may be formed well above the normal high tide range (Figure 4.35B). In the wave-breaking zone on the upper beach during moderate wave action there are often detailed ephemeral features such as beach cusps (Figure 4.35C) maintained by the backwash being channelled down the beach. Almost all beach sediment is highly mobile, but on warm tropical coasts with calcium carbonate dissolved in seawater, cementation of beach sediments can occur to form **beachrock**.

Other coastal sedimentary environments include estuaries and deltas where marine and fluvial zones interact and, above the water line, dune systems. Dunes are most common

Figure 4.35 Beach morphology. **A.** Sand spit forming a tombolo, linking an island to the mainland, near Denmark, Western Australia. **B.** Shingle storm beach, Porlock Bay, Somerset, England. **C.** Beach cusps on a shell beach, Shark Bay, Western Australia.

on low-lying coasts with a gentle seaward slope in the inter-tidal zone with a high tidal range. Large areas of sand exposed at low tide provide the source for sand to be blown onshore to form dunes.

4.6.4 Interactions with coastal biology

In the coastal environment there are important interactions between geomorphic and biological processes. On high-energy rocky coasts only algae can survive, the larger forms being anchored to the surface of the rocks. Algal crusts can also colonise rock surfaces. Sometimes these can play an important role in rock weathering. For example, on many limestone coasts in the Mediterranean and on many exposed coral platforms on tropical coasts, algae colonise the waterline and emit sufficient acid to counter the natural alkalinity of the seawater so that the limestone or coral rock (also composed of $CaCO_3$) is dissolved. The rough, pitted surface of exposed coral platforms has been produced by biologically-induced solution (**phytokarst**), and in many places what appear to be wave-cut notches are, at least in part, biochemically formed.

The most important interactions with ecosystems, though, occur on low-energy coasts. In sheltered environments (coastal lagoons, estuaries, less commonly on open, low-energy coasts) mudflats may be colonised by **salt-marsh** plants (Figure 4.36A), such as *spartina* or *salicornia*. Once these pioneer plants have begun to stabilise the mudflat environment a plant succession follows whereby other less tolerant plants colonise the marsh surface. The plants trap suspended sediment during high tide, building up the marsh platform, to create a complex micromorphology intimately related to the vegetation succession. That morphology includes the marsh platform itself, depressions in the platform, or pans that retain some of the tidal water and therefore retain a high salinity, and a system of muddy creeks that drain the marsh during the falling tide. Eventually the build-up of the marsh will

Figure 4.36 Coastal biology–geomorphology interactions. **A.** Saltmarshes, Grange-over-Sands, Cumbria, England. **B.** Mangrove, Queensland, Australia. **C.** Pleistocene coral rock that formed a fringing reef during the last interglacial marine highstand (OIS5), now exposed above modern sea level, Grand Cayman Island.

lead to much less frequent inundation, and a salt marsh becomes a brackish, or even a freshwater marsh. The equivalent in tropical areas is **mangrove** (Figure 4.36B), a woody bush that can root under salt water, but having a similar effect on sedimentation as do salt marsh plants on temperate coasts.

Another important interaction with biology, again restricted to tropical environments, is the development of **coral**. Modern coral coasts are found primarily in the Caribbean, The Red Sea, south-east Asia, Australia and the tropical Pacific islands. Corals are static colonial animals that build an exoskeleton by secreting calcium carbonate. They can only survive in warm, relatively clear seas with a pH above 7. If sea level rises they build colonies upwards and outward towards the sea surface. They die if sea level falls, leaving behind coral rock (Figure 4.36C), so that they are a very sensitive indicator of sea-level change. Modern live coral grew to its present position during the post-glacial global sea-level rise. Today it occurs primarily in three forms: **barrier reefs** forming offshore platforms; **fringing reefs** attached to shore; and **atolls** circular forms enclosing a lagoon, that grew initially as fringing reefs around an island that has since been submerged.

4.6.5 Coastal change

Inevitably, in some of the discussion above we have had to consider coastal change and evolution. The most important consideration is sea-level change. We have already considered the effects of the Pleistocene glacial sequence on global sea levels (Section 2.2.4). Modern sea levels date only from the last 6000 years or so. Modern coasts are young features, wholly Holocene in age. However, during previous interglacials global sea levels were also high, at or above modern sea levels. In places, remnants of former coastal features, raised wave-cut platforms and **raised beach** sediments (especially from the last full interglacial OIS5, 125,000 years ago; Section 2.2.4), are preserved above the modern coasts, for example at a number of sites in south-west England and in many places around the Mediterranean (Figure 4.37). Many tropical islands are formed of Pleistocene coral rock related to past higher sea levels.

The Holocene sea-level rise was complicated in some areas, giving some coasts an accentuated 'drowned' coastline, and on other coasts even creating 'emergent' coastlines. The main complications relate to either tectonic activity or to glacio-isostasy (Section 2.2.4). For example, tectonic downwarping of the Musandam peninsular on the Arabian Gulf has accentuated the drowned nature of that coastline; tectonic uplift of the central Californian coast has created an emergent coastline there. Glacio-isostatic depression followed by sustained isostatic uplift has created complex coasts with both submergent and emergent features in parts of the Arctic and North Atlantic Oceans, in western Scotland, Hudson's Bay, eastern Canada and around the Baltic Sea (*see* Figure 2.8).

Within the context of the last 6000 years, modern coasts are unlikely to have reached any sort of equilibrium condition, and are therefore subject to continuing change. The

Figure 4.37 Coastal change – raised beaches. **A.** Last interglacial (OIS5) raised pebble beach sediments, now cemented and perched *c*5 m above the modern shoreline, Carboneras, Almeria, Spain. **B.** Detailed view of sediments comprising sandy pebbles, rich in bivalve fossils, from last Interglacial (OIS5) raised beach deposits, Macenas, Almeria, Spain

mid-Holocene irregular submergent coastlines have been straightened by cliff erosion and by deposition in lower lying areas. On depositional coasts there has been a progressive evolution of depositional features such as spits, and sedimentary infilling of embayments. River deltas have grown and estuaries have been subject to infilling.

Overall, after glaciation there was an abundance of easily erodible sediment, followed by a decline in sediment availability, clearly accentuated in the last century or more by human activity (*see* Chapter 6), such as coastal protection schemes and, most importantly, by damming of many major rivers, trapping their sediment within the river basin and preventing its yield to the marine system.

A final consideration is the question of the potential response to the predicted *c.*2 m rise in global sea levels over the next century as a result of global warming.

4.6.6 Lake shorelines

Before leaving coastal geomorphology, we need to say a little about lakes and their shorelines. Lakes vary in size from the smallest pond to the huge bodies of water of the North American Great Lakes.

There is an important distinction between outlet lakes and terminal lakes, which is primarily a water-balance question. In humid areas excess water maintains lake levels to the level of the outlet, discharging water to a river system downstream, but in dry regions lake levels may not reach the basin outlet and will fluctuate in response to the short-term water balance. In arid areas this may mean that the lake basins are dry most of the time, except following exceptional precipitation. These lakes are saline, depositing salts on the (playa) lake floor (*see* Section 4.1.3). When dry they provide a source of salt-rich dust and sand for aeolian processes.

How do large lakes differ from the Ocean? They are generally not saline, though some dryland inland seas are. They do not have any tide to speak of. Many have existed only since the late Pleistocene (10,000–12,000 years), though in some areas large lakes are older, so most have shoreline features whose ages are comparable with marine shorelines. Shoreline processes on the largest lakes are wave-dominated, causing cliff erosion and beach development, but within a restricted vertical range. Input deltas are important features, and lake basins are subject to gradual infilling from the input zone. One feature that is important on the larger Canadian and Russian lakes is winter ice, influencing shoreline processes.

Permanent lakes act as local base levels for the input streams and rivers. Desert playas have an interesting base-level influence. Over time the lake floor is subject to sedimentation, raising the base level for any input streams, causing the deposition of a wedge of fan-delta sediments on the margins of the playa. On the other hand, the large 'pluvial' lakes that occupied some of the larger interior basins of the Basin and Range region of the American West (e.g. Pluvial Lakes Bonneville, Utah, and Lahontan, Nevada) during the late Pleistocene not only provided a raised base level at the basin margins, but created a local base-level fall as the lakes desiccated at the end of the Pleistocene. Where this coincided with a steep shoreface, incision of input streams was triggered; elsewhere the deposition of sediment from these streams simply shifted towards the centre of the basin.

5   Timescales and landscape evolution

When it comes to explaining landscapes, there is more to consider than simply explaining individual landforms or even landform suites. We can explain landforms by considering processes and process–form relationships within the context provided by the setting and the geology. To explain landscapes we need to consider the relationships between landform suites and how they have changed over TIME. In other words we need to return to some of the issues raised by temporal and spatial scales in Chapter 1. To do this we will start by considering an example.

5.1   Landscape evolution: an example – late Quaternary landscapes of Carlingill, Cumbria.

Figure 5.1 shows part of the Carlingill valley in the Howgill Fells of north-west England. The photo shows details of the landscape, hillslopes, gully systems both active and inactive, streams and their deposits, and a large, stony fan-like feature in the centre of the photo. At this scale we are dealing with a landscape that has evolved over a relatively short period of time. The longer term merely provides the context. The bedrock geology is of folded Silurian mudstones, and the area is tectonically stable. The hilltops are part of an ancient dissected erosion surface. We know from the regional setting that during the Last Glacial Maximum (about 20,000 years ago) this part of northern England was under a considerable thickness of glacial ice, so we are essentially dealing with forms that have evolved

since the decay of that ice sheet, estimated to have occurred around 15,000 years ago.

Bedrock outcrops on the upper part of the hillslope to the right of the photo, but the lower part of the slope is composed of glacially derived boulder clay, stones set in a clayey matrix. Exposures at the bottom of the slope, where it is cut by the stream, show a very compact deposit at the base. This is probably a lodgement till deposited as the ground moraine of the ice sheet. The ice overrode the ridge behind and above the photo scene and was banked up against the bedrock of the valley side from which the photo was taken. The upper parts of the same sections show a less compact deposit with a weak stratification and stone alignment parallel with the hillslope surface above. This deposit has clearly been emplaced by moving downslope, the product of solifluction, reworking the glacial till. Solifluction also created the smooth, gently concave hillslope, almost certainly during the period after the decay of the ice sheet before the end of cold (permafrost?) conditions at the end of the Pleistocene.

The solifluction surface can be traced around the hillslopes. Its lower edge is terminated by a steeply undercut slope, clearly produced by later stream incision. On the left-hand side of the photo the equivalent surface grades gently to a small flat area that coincides with a coarse, gravelly deposit resting on a horizontal surface cut across the glacial till. This is a stream terrace some way above the

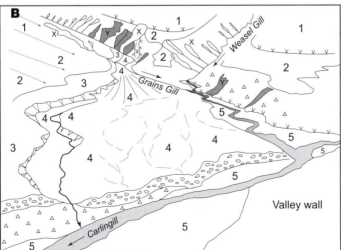

Figure 5.1 Grains Gill, Howgill Fells, Cumbria, England. **A.** View of Grains Gill catchment, tributary to Carlingill (foreground, flow from right to left). **B.** Interpretive sketch of the scene depicted in Figure 5.1A. **Key:** *1.* Bedrock hillslope; *2.* Late Solifluction surface; *3.* High stream terrace and fan surface; *4.* Lower fan surface (forms the lower terrace above Carlingill; *5.* Modern stream sediments, forming the floodplains.

Y   Active gullies and scars
X   Older (now stable) gullies
1 - 5   Erosional and depositional land sequences

Stream/fan gravels
Boulder clay

modern stream, and probably formed during the same timespan as the solifluction surface. Its continuation can be traced as a bench on the left-hand side of the side valley.

All the ground below this terrace is cut into it, and is therefore related to incision into the terrace, and is therefore younger. On the hillslopes and cut into the solifluction surface are a number of now-stabilised hillslope gullies. These grade below the level of the stream terrace into the stony fan feature in the centre of the photo. On the fan we can see bouldery, lobate ridges and the traces of a braided channel system. The gully–fan association suggests a major period of hillslope erosion and fan deposition. At the time sediment supply must have been greater than the stream system could have transported away. The modern stream system is set below the fan surface, suggesting that as the gullies stabilised, sediment excess conditions were replaced by conditions under which again the stream could incise. The modern gullies and streamside scars are active today and feed sediment into the modern stream channel.

By examining the landscape and considering the relationships between landforms and sediments, we have been able to construct a sequence of events that have produced the landscape we see. In brief, that sequence was as follows: i) deglaciation, followed by the formation of the solifluction surface; at the same time the stream terrace was formed; ii) a period dominated by incision, during which time the terrace was abandoned and a valley was cut below the terrace and the solifluction surface; iii) a period of hillslope gullying and deposition of a large alluvial fan on the valley floor; iv) stabilisation of the hillslope gullies, reduction of sediment supply and the incision

of the modern stream below the fan surface; v) modern gullying, but not producing sufficient sediment to overload the system; the sediment being fed into the modern stream. Apart from suggesting a crude timescale relating to deglaciation, and later to the cessation of solifluction, what we have not yet done here is to date some of these events, and convert our landform/sediment sequence into a chronology.

5.2 Relative dating

In the example above we have used evidence from the landforms themselves and their relation to sediments to suggest a sequence of development. There are two other types of evidence we can use to refine such a sequence: what we can call relative and absolute dating. Relative dating depends on recognising a timescale that can be applied to a landform/sediment sequence.

The first group of relative dating techniques deals primarily with what is contained within a body of sediment that constitutes a depositional landform, the equivalent of fossils in sedimentary rocks. Conventional fossil evidence, based on the evolution of life forms, is of little use in late Pleistocene and Holocene sediments, although fossils do have a role in dating older Pleistocene deposits. For example, raised beach deposits in the Mediterranean contain a tropical gastropod fauna not present in the Mediterranean today. The sediments containing such fauna cannot be of late Pleistocene or Holocene age, and must date from the last interglacial (OIS5) or before. The most commonly used fossil evidence from late Quaternary deposits is fossil pollen. Pollen is best preserved in peaty sediments or buried organic horizons, and has to be analysed in the laboratory. The species contained in a fossil pollen assemblage may help ascribe a sediment to a known part of the late Quaternary vegetation sequence including, from the mid-Holocene onwards, to known periods of human impact on the vegetation sequence. In the arid areas of the American southwest pollen is preserved, cemented into packrat nests by rat urine. Another particular type of 'fossil' evidence applicable to the mid-Holocene onwards is archaeological evidence in the form of human artefacts.

More important, perhaps, because the evidence is much more widespread, is the modification of depositional surfaces by processes for which we have some knowledge of the rates of development. The classic case is soil formation. In the Howgill example above (Section 5.1) the solifluction surface and the high stream terrace are both characterised by well-developed **podzolic** soils with a red-brown B horizon, whereas the younger surfaces at best have only a weak A–C profile of an organic horizon over barely weathered parent gravels. In other words, we can identify a soils **chronosequence**. Many lichens (Figure 5.2A) have known growth rates and have

Fig. 5.2 Dating of Holocene landforms in the Howgill Fells, Cumbria, England. **A.** Lichen mosaic colonising a boulder. Lichens such as these are suitable for dating valley floor bar forms of ages back through the last 200 years or so. **B.** A small alluvial fan/debris cone, tributary to Carlingill. Note the higher stream terrace (of late Pleistocene age), cut through by multi-segmented fan, the upper segment of which grades into the lower terrace of Carlingill. Note the mobile channel of Carlingill on the main valley floor (flow from left to right); previous courses of the stream are marked by cobble bars on the valley floor. **C.** Buried organic soils (the dark layers) interstratified with fan deposits, suitable for radiocarbon dating of the phases of fan sedimentation. **D.** Summary diagram (produced by Richard Chiverrell), of alluvial fan sedimentation phases

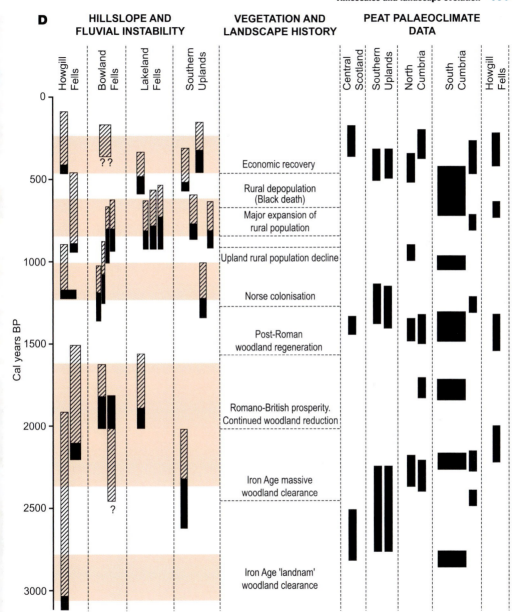

D HILLSLOPE AND FLUVIAL INSTABILITY

VEGETATION AND LANDSCAPE HISTORY

PEAT PALAEOCLIMATE DATA

dated by radiocarbon dating, for upland areas in northwest England and southwest Scotland. The main phases of hillslope activity (shown on the left of the diagram by pink bars) appear to correlate primarily with human impact on hillslope vegetation, rather than with climatic wettening periods (black bars, right hand side of the diagram).

been used in many areas to date depositional surfaces, such as recessional moraines in areas of modern glaciation. In the Howgills we have used **lichenometry** to correlate depositional surfaces within the modern stream system. The succession of higher plants may also be useful if something is known of the rates of vegetational succession. Perhaps the most useful form of such biological evidence would be tree ring counts (**dendrochronology**) from trees growing on depositional surfaces. Such evidence (including distortion of tree ring growth) has also been used to assess the frequency of debris flows in the Alps and other mountain areas.

Most, but not all, of the applications above have been applied primarily in humid areas. There are similar techniques applicable to arid environments. These include the following: the degree of desert pavement development (Section 4.1.1, Figure 4.1B); the degree of desert soil development (Section 4.1.3; Figure 4.3C), and the maturity of calcrete development (Section 4.1.3; Figure 4.4A), for example whether the calcrete is simple or shows evidence of **brecciation**. In addition, applicable also to the exposure of rock surfaces, is the degree of development of **desert varnish**, a dark iron and manganese coating that develops in arid environments.

There are far fewer techniques available for relative dating exposure of rock surfaces than of depositional surfaces. Lichens, weathering rind or weathering pit development are those most commonly used.

5.3 Absolute dating

The relative dating techniques outlined above may help to refine a sequence, but in few cases give absolute dates. There are, however, a number of analytical techniques that can give absolute dates to a sequence. All depend on sophisticated laboratory procedures. Without getting too technical, it is necessary here to give an outline of the applicability of some of the more important techniques used. Most involve the dating of sediments, so field sampling is important to establish whether the resulting date will relate to before, during or after the event being dated. Many of the techniques depend on the radioactive decay of isotopes within minerals that form the constituents of sediments. The applicability depends on the half-life of radioactive decay of the particular element involved; for example, potassium–argon series dating is appropriate for timescales of millions of years and has been more important in dating rocks than more recent sediments.

The most commonly used method in dating sediments is **radiocarbon dating**, measuring the proportion of ^{14}C to ^{12}C ions in organic (peat, organic soil, wood, charcoal) or formerly living carbonate sediments (marine shells). The standard technique is applicable to the last 30,000 years or so, and therefore has been extremely useful for dating late Quaternary sediments. It has also been invaluable in archaeology. The standard technique has been in use for the last 40 years, and depended on relatively large amounts of carbon. More recently with the development of the **AMS** (Accelerator Mass Spectrometry) method much smaller samples can be used, extending the applicability of the method, both in general terms and in the dating timescale.

Another important radiometric method that has been developed over the last 20 years depends on the uranium–thorium series, found in trace amounts in calcium carbonate.

This method has an applicability of several hundreds of thousands of years, so has been useful in dating carbonates (marine shells, pedogenic carbonate) from the middle and late Pleistocene.

Two other radiometric methods deserve mention, both applicable to very recent sediments – the lead 210 and caesium 137 methods. Caesium 137 is an artificial element, generated by nuclear explosions. The method can therefore only be used to date very young sediments, deposited since the late 1940s, and to estimate rates of recent soil erosion or, for example, floodplain deposition. Lead 210 dating is applicable to relatively young sediments, from the last hundred years or so.

Two other recently developed dating techniques, which do not depend on radioactive decay, have been applied to geomorphological questions: **luminescence dating** and **cosmogenic dating**. Quartz or feldspar crystals within buried sediments absorb background ionising radiation, which, with stimulation, can be released as luminescence. The luminescence is bleached away by exposure to sunlight. The luminescence signal, which increases with burial time, can be used to date the last exposure of the sediment to sunlight. This method has been successfully applied to dating quartz- or feldspar-rich sandy sediments, particularly from aeolian or fluvial environments, environments where exposure to light occurs during transport, followed by rapid burial on deposition. It is applicable for late Quaternary timescales of the last 100,000 years or so. In contrast with most of the radiometric decay methods available over these timescales, it gives a date of sedimentation, rather than a before or after date. Cosmogenic dating measures the absorption of cosmogenically derived nuclides onto exposed rock surfaces, so it has the potential for dating rock exposure.

5.4 Temporal and spatial scales – synthesis: two examples

We will conclude this chapter by considering landscape evolution in two contrasting areas, where conventional morphostratigraphic relationships, augmented by relative dating methods, have been given more precision in recent years by the application of more advanced absolute dating methods. The two examples stem from my own research, but have involved cooperation with colleagues, especially in the context of dating. The two examples relate to very different geomorphologies and cover very different timescales.

5.4.1 Upland north-west England

The first example builds on the Howgill example developed above (Section 5.1). There we identified now stabilised hillslope gullies that fed sediment into an alluvial fan system. Similar gully systems throughout the Howgill Fells and other hill areas in north-west England and south-west Scotland fed sediment to debris cones and alluvial fans (Figure 5.2B). The first question to ask is how old are these systems? The morphological and soil evidence suggests that they are Holocene, not late Pleistocene, in age. The second question is whether they represent only one or several periods of slope instability, and whether they were synchronous throughout the broader region. The third question is, what was or were the likely cause(s) – climatic fluctuations or human impact on the landscape? We might hope to make some suggestions in relation to this last question if we knew more about the timing of slope instability.

Sections cut by stream erosion into the toes of some of the fans and debris cones expose either buried soils or soil horizons within the fan/cone sediments (Figure 5.2C). The organic horizons of some of these soils, initially primarily from sites within the Howgills, were sampled and dated by standard radiocarbon dating. Later, particularly through cooperation with Richard Chiverrell of the University of Liverpool, we extended the cover to other upland areas within the region, and used many more samples, including some that involved AMS dating.

The results are summarised on Figure 5.2D, and throw light on the three questions posed above. First of all, almost all the dates relate to the late Holocene, and suggest that the soils became buried by fan/cone deposition during the last 2000 years. Secondly, there appear to have been several phases of slope instability. Although there are subtle within-region differences, there is overwhelming evidence for a major period(s) of activity between 1000 and 700 years ago. There was another less widespread period during the last 300 years. On examining the bunching of dates and relating it to evidence of climatic fluctuations and of human impact, the data strongly suggest that the main phase(s) of slope instability relate primarily to human impact, through the effects of grazing on the uplands and the associated degeneration of an earlier woodland cover. On the other hand, the last phase of slope instability, initiated during the last 300 years, might have been influenced by the seventeenth- to nineteenth-century climatic deterioration known as the 'Little Ice Age'.

5.4.2 The Sorbas basin: south-east Spain

The second example relates to an area very different geomorphologically, and spans a much longer timescale than the above example from north-west England. It deals with the Sorbas basin, a small sedimentary basin within the Betic cordillera in south-east Spain over timescales of the last million years or so. First, we need some longer-term geological context. The region underwent mountain building as a result of the collision between the African and European tectonic plates, culminating in the mid-Miocene. By then the major basement structures had been emplaced, which now form the mountain ranges of the Sierra de los Filabres and Sierras Alhamilla/Cabrera (Figure 5.3) and the intervening sedimentary basins, of which the Sorbas basin is one.

Since then, neotectonic activity has continued, and continues to the present day, expressed in the development of a major strike-slip fault system that partially determines the gross relief configuration. In addition there has been sustained regional differential uplift, accentuating the regional relief.

During the late Miocene sedimentary basin filling occurred, initially as marine sedimentation (including the deposition of gypsum deposited as the Mediterranean became landlocked and desiccated). By the early Pliocene the Sorbas basin became emergent and underwent terrestrial sedimentation, but the Almeria basin to the south (Figure 5.3) remained a shallow marine environment. By the end of the Pliocene/early Pleistocene a river system (the Aguas/Feos) had developed from the Sierra de los Filabres south across the Sorbas basin and into the Almeria basin, across a structural low between the Sierras Alhamilla and Cabrera. This drainage persisted for much of the Pleistocene. A series of river terraces, composed of fluvial gravels and conglomerates, can be traced through the Sorbas basin and into the Almeria basin that represents stages in the incisional drainage development (Figure 5.4A).

Meanwhile, another river system (the lower Aguas), draining to the east into the tectonically lower neighbouring Vera basin (Figure 5.3) was cutting back as a subsequent drainage (*see* Section 3.2) along the outcrop of a particularly weak marl within the upper Miocene marine sediments. It cut back into the centre of the Sorbas basin and captured the original Aguas/Feos drainage. This river capture was to have huge implications for the geomorphology of the Sorbas basin.

The author, with colleagues Steve Wells, Anne Mather, Martin Stokes and Elizabeth Whitfield (Maher), identified the response to the capture. The upper Aguas rapidly incised back into the Sorbas basin, creating **canyons** and **incised meanders** (Figure 5.4B), triggering landslides and initiating slope erosion and badland development (*see* also Figures 4.6C, 4.9C). The beheaded Feos lost all its power and became an underfit stream, incapable of further incision (Figure 5.4C, D). Downstream the Feos joins the Alias, which was transformed from a large to a much smaller river, by the loss of the Aguas/Feos headwaters.

To give us some idea of the rates of these geomorphic processes we need to be able to

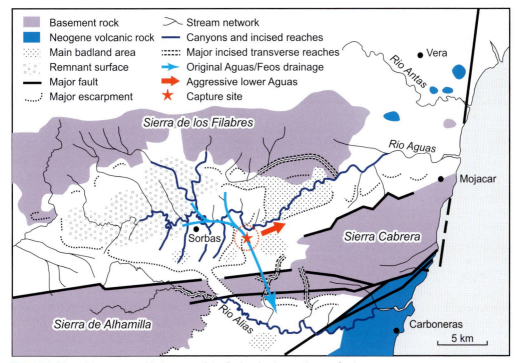

Figure 5.3 Map of the geomorphic features of the Sorbas basin, south-east Spain.

A

B

C

D

Figure 5.4 The Sorbas (SE Spain) river capture event. **A.** The river capture site, looking south into the beheaded and now abandoned Aguas/Feos valley. Flow was southwards, away from the camera, through the perched valley in the centre of the photo. That valley floor is lined with former fluvial gravels. That drainage was captured by the lower Aguas, which now flows east, right to left, within the incised valley in the centre of the photo. Note how the rapid incision has triggered slope instability by undermining the limestone slope (left centre of the photo). **B.** Evidence for the timing of the river capture was based on an examination of the river terraces of the former river system. Here two gravel terraces of the former Aguas/Feos rest unconformably on late Miocene marine sedimentary rocks (upstream of Sorbas village). The highest terrace forms the skyline on the right of the photo. The next terrace down forms the bench on the left side of the photo. **C.** The effects of the capture upstream from the capture site: Incision worked its way upstream from the capture site to cause spectacular incised menders at Sorbas village. Note: the two flat surfaces on which the village is built are river terraces of the former Aguas/Feos river. **D.** The effects of the capture downstream of the capture site on the beheaded Rambla de los Feos. The shrunken river has not incised below the former valley floor. Instead the valley floor, where it passes through the Sierras Alhamilla/Cabrera is almost choked by small tributary alluvial fan and colluvial hillslope sediments.

date the capture event. First, using relative dating, and working with Steve Wells and Suzanne Hunter (Miller) on the soils and calcrete development on the river terraces deposited before and after the capture, we estimated that the capture took place between

50,000 and 100,000 years ago, with the higher terraces being considerably older and the post-capture terraces in the Feos valley being considerably younger, the youngest of which we had dated by radiocarbon dating to the Holocene (2300 BP +/- 90 a).

More recently Ian Candy, using uranium–thorium dating of calcrete and pedogenic carbonate, has confirmed a date of approximately 60,000 years ago for the capture event, giving a relatively short timescale for the major geomorphic changes that followed it. Using luminescence dating of the sediments within the main post-capture terrace in the Feos/Alias valley, Andreas Lang, Barbara Mauz and Elizabeth Whitfield have confirmed a date of around 20,000 years, a date that confirms the increased sediment yields coincident with climatic changes associated with the last global glaciation (OIS2). This again gives a timescale for the development of the younger landforms in that valley.

6 Geomorphology and interactions with society

In this final chapter we consider interactions between geomorphology and society. This is a huge topic; in this short chapter we can only scratch the surface.

6.1 Human impact on geomorphic regimes

Over the last 5000 years or so human societies have modified the natural environment, especially since the Neolithic revolution and the widespread development of agriculture. In many parts of the world the natural vegetation has been radically altered by forest removal, grazing animals and crop cultivation, modifying the surface hydrology/run-off/erosion relationships. Initially this took place in the vulnerable semi-arid environments of the Middle East and the Mediterranean region. Much later, over the last 2000 years, woodland clearance in Europe also replaced the natural vegetation by grazing and cropland. The geomorphic impact was to accelerate slope processes, for example, leading to extensive land degradation in the Middle East and in the Mediterranean, to hillslope gullying in upland Britain (*see* Section 5.4.1), and to slope instability in some of the steeper loess terrain of southern Germany. The impact was also felt further down the sediment cascade, on the stream systems. Extensive valley-floor aggradation, and later dissection, led to the formation of late Holocene, human-impact river terraces throughout the Mediterranean world, and to a lesser extent in many areas of Europe.

The evidence for the causes of these changes is inextricably linked with archaeological and palaeovegetational evidence, but the evidence for the geomorphic impact is contained in the landforms and sediments themselves.

Over the last 200–300 years the impact of agriculture has extended to the New World, the Americas, Australasia and other areas of European colonisation in Africa and Asia. Extensive agriculture has led to widespread soil erosion throughout the world, but perhaps the biggest human impact has occurred in the last 200 years with the developments of mining, industry, urbanisation and tourism, to mention only the most obvious changes in land use. The effects include not only indirect responses of the sediment system to land-cover change, but also the direct engineering manipulation of the physical environment.

River systems have perhaps undergone the most radical changes. Most of the largest river systems have been dammed for various purposes – irrigation, water supply, hydroelectric power and river regulation. The effects have been to trap most of the river sediment in an artificial reservoir behind the dam, thus reducing the sediment flux downstream, and ultimately to reduce the sediment yield from the continents to the oceans, despite the order of magnitude increase in sediment production from the land surface. There are local effects too. Downstream from dams the water is clear, so that despite a common reduction in the magnitude of flood peaks, erosive power is enhanced. There have been many documented cases of river incision downstream from dams. Many rivers, too, have been channelised as a defence against erosion and local flooding. However, the effects have often been to increase flood power and to transfer a flood problem downstream.

The enormous scale of modern human intervention is nowhere better illustrated than by the Three Gorges Project on the Yangtze River in China. The dam itself is 2.3 km wide and 180 m above the rock base (110 m high above the downstream river level). It impounds a lake within the Three Gorges reach of the river that will extend more than 600 km upstream and have a shoreline over 2000 km long. The purpose of the project is three-fold: to generate power for the rapidly developing Chinese economy; to alleviate the chronic and disastrous flood problems in the densely populated middle reaches of the Yangtze basin; and to improve navigation upstream to Chongquin. In the long run it may also allow the transfer of water to areas of chronic water shortage in northern China. Whether these aims will be fully satisfied remains to be seen. In addition to the human impact of moving the 1.5 million people directly affected by the project, the impact on the geomorphology will be severe. Sediment will be trapped in the lake. The lake shoreline zones within the steep slopes of the Yangtze Gorges have already had catastrophic landslides. The whole fluvial system for the approximately 2000 km downstream to the delta will be affected by change

as the system adjusts to the major changes in flood power and sediment flux that will follow.

In many parts of the world, the coastal environment, too, has been transformed by engineering works, beach groynes, sea walls and other coastal defences. The effects have often been to starve parts of the coast of sediment, resulting in beach erosion, and in some cases causing renewed cliff erosion.

Today, in the early years of the twenty-first century, the greatest potential human-induced environmental change is global warming, with the prospect of affecting all environmental systems, including geomorphic systems. The direct effects of climate change may include changed temperature regimes, which could affect many geomorphic processes. Already sea ice and glacier retreat in high latitudes are taking place at an alarming rate. Almost all mountain glaciers are also in retreat. Thawing permafrost could affect surface and slope stability in the sub-arctic. More significant, perhaps, could be changes in precipitation regimes, by an increased incidence of drought and potentially increased storm rain intensity, both of which could affect slope and fluvial processes. The predicted effects of global warming include the possible increase in the incidence and intensity of hurricanes. There could also be intensification of the 'El Niño' and 'La Niña' oscillations in ocean circulation in the Pacific, with its associated effects on precipitation patterns in Central and South America and the south-east Asia/ Australasia region. On the world's seacoasts rising sea levels predicted as a consequence of global warming could be devastating. Apart from the potential inundation of low-lying areas and the obvious human consequences, even a small rise in sea level will transform many coasts into erosional coasts. To understand and hopefully to mitigate the potential effects of global warming, environmental planning will need to take into account the vulnerability of all geomorphic systems.

6.2 Applied geomorphology

Geomorphology interacts with society in a number of ways, particularly in relation to hazards, resource management and environmental conservation, so that some aspects of geomorphological research are directed towards these issues. In addition to geologically induced hazards, such as earthquakes and volcanic activity, hazards induced by geomorphological processes are significant. These include flood hazards, globally the most severe environmental hazard. Though primarily a hydrological hazard, mitigating its potential impact often depends on understanding fluvial geomorphology. The flood hazard is particularly severe on large lowland rivers bounded by extensive floodplains, but there are also severe hazards in mountain areas associated with debris-flow floods running out onto debris cones or alluvial fans. Over the last 20 years there have been a number of such disasters in the Alps and the Pyrenees. Mountain areas may also be prone to larger scale landslides, particularly where vegetation removal has increased the potential for landslide activity. The effects can be devastating, especially where inappropriate land uses are situated in landslide-prone sites. Much Italian geomorphological research has been directed to assessing landslide risk in mountain areas. Although not a short-term hazard, soil erosion is a severe global environmental problem, and the subject of a large field of geomorphological research.

Resource and environmental management are fields where the application of geomorphological knowledge is important. Effective coastal management depends on an understanding of coastal geomorphology. The assessment of sand and gravel resources for the construction industry often depends on understanding fluvial sedimentation history, or in many areas of northern Europe or North America, fluvioglacial processes. It is not only in such resource assessment that geomorphological knowledge has a role, but in assessing the impact of extraction on the local environment. Indeed, in most countries of the Western world, environmental impact assessment is an essential prerequisite for resource exploitation, whether it be for mining or for general and urban development.

Another planning field in which geomorphology has a role is in designating and managing environmental conservation areas. Whether these are essentially wilderness National Parks, as in the USA, or conservation sites of specific biological, geological or geomorphological importance, understanding the geomorphology is important for effective conservation of the natural environment.

Although many aspects of applied geomorphology are implemented by government agencies, often employing engineers or planners, much of the fundamental research is undertaken by geomorphologists.

6.3 Geomorphology in education and research

Understanding geomorphology is important not only as a science in itself but, as indicated above, is relevant to the effective management of the environment and its resources. Education in geomorphology therefore has a role to play. As indicated in Chapter 1, geomorphology as an academic discipline sits between the disciplines of geography and geology. In most European countries geomorphological teaching, both at school and university levels, is primarily in the context of geography. At university level in the USA, geomorphological teaching has a greater emphasis in geology courses, but does have a presence in geography.

This pattern is also apparent at the research level. In Europe, especially in Britain, France, Germany and the Scandinavian countries, most university-based research in geomorphology is carried out within geography or physical geography departments, mostly funded by government grants of some sort. In countries that inherited a system based on the British system, notably Australia, New Zealand and Canada, the situation is similar. There are of course variations. In Spanish universities there is both geography- and geology-based geomorphological research. In Italian universities engineers and geologists dominate geomorphological research. In the USA geology departments dominate the field, but there is also significant research output from some geography departments. In some countries government agencies undertake primary research in geomorphology and related fields, sometimes by scientists employed by government, sometimes by contract with university scientists. For example, in the USA the USGS (United States Geological Survey) is important in this area; in France the CNRS (Committee National de Recherches Scientifiques) has such a role, as does the CSIRO (Commonwealth Scientific and Industrial Research Organisation) in Australia.

When it comes to applied research and environmental management, this is mainly done by government agencies, less often by private consultancy firms. Both draw their talent from academic fields, particularly from engineering or from geology, but whose work is dependent on fundamental research in geomorphology.

Geomorphological research is reported in a wide range of scientific journals, some of them subject-based, particularly in geology, some of them interdisciplinary. Two specifically geomorphological international journals deserve mention: *Earth Processes and Landforms* and *Geomorphology*. There are also learned societies and professional bodies that promote geomorphology. The International Association of Geomorphologists is an umbrella organisation, linking national societies, of which the British Society for Geomorphology (formerly the British Geomorphological Research Group), which is affiliated with both the Royal Geographical Society and the Geological Society, is the oldest and one of the largest. In the USA there are geomorphology sections within both the Geological Society of America and the Association of American Geographers. The American Geophysical Union is also involved with geomorphology. In Europe, in addition to the various national organisations, geomorphology is a major component of the recently formed European Geophysical Union.

The present picture is not static. Geomorphology as a scientific discipline is continuously developing, and university organisation also changes, as does perception of the environment by the general public. Within the subject there is a growing need for interdisciplinarity. There are new technologies developing that affect the training of scientists. For example, there are developments in remote sensing and in the application of GIS

(Geographical Information Systems) and computer-based modelling that have important implications for geomorphological research. There are new technologies in environmental monitoring, not to mention the development of advanced, sophisticated dating techniques that all have implications for geomorphological research. These need to be applied alongside traditional field-based research.

Within universities the emergence of interdisciplinary schools of, for example, Earth Sciences or Environmental Sciences, amalgamating former single subject departments, should be beneficial to geomorphology. Funding for research is also changing, with a growing emphasis on applied research. Given the increased public awareness of the environment, this may not be entirely a bad thing! Above all, the present picture is dynamic; geomorphology is a healthy subject, with a broad international base and a growing role in understanding our natural environment.

Further reading

This book is not intended as a textbook; there are a large number of textbooks available on geomorphology at all levels from the high school through university level textbooks to research texts. Some of them are 'classics' written more than 50 years ago, but still useful. At university level there are numerous 'thematic' texts that deal with sub-fields of geomorphology (e.g. fluvial, glacial, tropical, coastal geomorphology, etc.). A very personal selection of those that I find most useful includes the following.

Classics

Thornbury, W.D. 1954. *Principles of Geomorphology*, Wiley, New York (2nd edn, 1969): a classic American text.

Leopold, L.B., Wolman, M.G. and Miller, J.P. 1964. *Fluvial Processes in Geomorphology*, Freeman, San Francisco: a classic advanced text in fluvial geomorphology, that fundamentally changed approaches to the subject.

Flint, R.F. 1971. *Glacial and Quaternary Geology*, Wiley, New York: the classic glacial text.

High school texts

Hilton, K. 1979. *Process and Pattern in Physical Geography*, University Tutorial Press: a basic English school text.

Strahler, A.N. 1975. *Physical Geography* (4th edn) Wiley, New York: THE basic American text!

University level texts

Ahnert, F. 1996. *Introduction to Geomorphology*, Arnold, London: a comprehensive, university level text, with an emphasis on geomorphic systems.

Summerfield, M A. 1991. *Global Geomorphology*, Pearson Educational, for Prentice Hall, Harlow: a comprehensive text, particularly useful for its global scale approach.

Encyclopaedic works

Goudie, A.S. (ed.). 2004. *The Encyclopaedia of Geomorphology*, Routledge, London: A comprehensive encyclopaedia, produced in association with the International Association of Geomorphologists.

Specialist fields (selected examples only)

Bird, E.F.C. 1984. *Coasts: an Introduction to Coastal Geomorphology*, Blackwell, Oxford.

Bull, W.B. 1991. *Geomorphic Response to Climatic Change*, Oxford University Press, Oxford.

Keller, E.A. and Pinter, N. 2002. *Active Tectonics: Earthquakes, Uplift and Landscape*, Prentice Hall, New Jersey.

Knighton, D. 1984. *Fluvial Forms and Processes*, Edward Arnold, London.

Lewin, J. (ed.). 1981. *British Rivers*, Allen and Unwin, Hemel Hempstead.

Schumm, S.A. 1977. *The Fluvial System*, Wiley, New York.

Selby, M.J. 1982. *Hillslope Materials and Processes*, Oxford University Press, Oxford.

Sugden, D.E. and John, B.S. 1976. *Glaciers and Landscape*, Edward Arnold, London.

Thomas, D.S.G. (ed.). 2011. *Arid Zone Geomorphology* (3rd edn), Wiley-Blackwell, Chichester.

Thomas, M.F. 1994. *Geomorphology in the Tropics*, Wiley, Chichester.

Useful websites

www.geomorphology.org – *The website of the British Society for Geomorphology.*

www.geosociety.org –*The website of the Geological Society of America, plus the useful sub-site of the American Quaternary Association* (www.rock.society.org)

www.qra.org.uk – *The website of the Quaternary Research Association (of the UK): contains information on the QRA field guides.*

www.usgs.gov – *The official site of the US Geological Survey – includes many useful sub-sites relating to such topics as: impact of climatic and land-use change; natural hazards; remote sensing; water.*

www.nps.gov – *The official site of the US National Parks Service – includes a useful sub-site on geological resources, with information on the geomorphology of the US National Parks.*

www.geoscape.nrcan.gc.ca – *A Canadian government website with information on Canada's North, climate change, natural hazards (including flood geomorphology, and landslides).*

www.ga.gov.au – *An Australian government website with information on hazards (including floods) and marine and coastal environments.*

There are numerous websites that present either photographs or satellite images of landforms at a variety of scales. Three are suggested below:

www.disc.sci.gsfc.nasa.gov – *A NASA collection of satellite images*

www.uvm.edu/~geomorph/gallery – *A superb and huge collection of landform photos organised by theme.*

www.flickr.com/groups/10716070@N20/ – *A good collection of photos, primarily of fluvial landforms.*

Also, do not forget Google Earth, a superb resource of satellite imagery covering the whole of the Earth.

Glossary

A

ablation [61]: the loss of glacial ice, primarily through melting on the glacier surface, but includes internal and basal melting and sublimation of ice directly to water vapour.

abyssal plain [16]: the majority of the ocean floor: forming a relatively smooth plain, on average 3.5 km deep, away from continental shelves, ocean trenches and volcanic hot spots.

active layer [51, 62]: the layer above permafrost that seasonally thaws, that becomes prone to mass movement processes, and on refreezing in winter produces the differential internal stresses responsible for the formation of patterned ground.

aeolian [79]: as in aeolian processes: erosion, transport and deposition of sediment by the wind.

aeolianite [80]: cemented dune sand, usually cemented by CaCO3, and occurring mostly in warm dry regions, often fossilising former coastal dunes.

aggrade/Aggradation [66]: the net deposition of sediment, resulting in an increase in elevation of the depositional surface. Applied particularly to river channels, floodplains and alluvial fans.

alluvial channel [69]: a river channel whose margins comprise alluvium (i.e. previously deposited river sediment). Processes in such channels usually involve a balance between erosion and deposition.

alluvial fan [75]: a conical or sub-conical depositional landform, deposited where a steep stream leaves the confinement of a mountain catchment, either at a mountain-front or a tributary-junction setting. Common in arid mountain areas, but occur under virtually all climatic conditions. Include a range of sizes from tens of metres to tens of kilometres in length. Sedimentation processes range from debris flows to fluvial processes, either as sheetflows or as streamflows.

alluvium [70]: sediment deposited by a stream or a river. Typically implies fine sediments, e.g. silts, but strictly includes all grain sizes.

AMS (Accelerator Mass Spectrometry) [102]: *see* **radiocarbon dating**.

anabranching [73]: a general term to describe all types of multi-thread river channels including both anastomosing and braided channels, though sometimes used interchangeably with 'anastomosing' channels.

anastomosing river channel [73]: a multi-channel river pattern, comprising relatively stable sub-channels separated by islands that are usually large in relation to the channel. A low energy form of anabranching channels, characteristic of mud-dominated, sediment-rich rivers, distinct from high-energy, coarse sediment, braided channels.

angle of incipient movement [58]: angle of a hillslope at the threshold of movement of clasts derived from upslope or produced *in situ* by mechanical weathering, somewhat steeper than the angle of rest of the material.

angle of rest [56]: depositional slope angle at which clasts of a given size come to rest.

antecedent drainage [40]: drainage lines that cross active fold structures, most obviously anticlinal folds, such that the rate of fluvial incision keeps pace with the rate of uplift, resulting in a drainage pattern transverse to structure.

anticline [37]: a fold in bedded rocks in the form of an arch. The upper (younger) rocks in the centre may be removed by erosion, exposing the older rocks underneath in the form of a breached anticline.

Arête [86]: A knife-edged ridge produced by the intersection the steep back or sidewalls of two glacial cirques.

arroyo [60]: a Spanish word simply meaning 'stream', but in geomorphology has a very specific meaning – an ephemeral entrenched channel that may be discontinuous; in other words, a valley-floor linear gully, characteristic of many areas in the American south-west. There is debate on the extent to which overgrazing, climatic change, or intrinsic instability may contribute to the causes of arroyo development.

atoll [94]: a circular (in plan view) coral reef, formed around an island or seamount, which has since been submerged by rising relative sea level.

avulsion [73]: river channel change by diversion and spillage during flood conditions into a backswamp environment adjacent to a river channel or onto a floodplain or alluvial fan surface, often causing the original channel to be abandoned. The new channel may rejoin the original channel further downstream, or, as is especially the case on alluvial fans, may represent a completely new direction of drainage.

axial channel [76]: as applied to channels on alluvial fans: the channel carrying water and sediment from the feeder catchment through the alluvial fan, usually runs down the centre or axis of the fan. Can also be applied to the 'axial channel' of a sedimentary basin, in which case it refers to a (main) channel running along the axis of a linear sedimentary basin, as opposed to tributary channels running from the basin sides into the centre of the basin.

B

badlands [59]**:** intensely gullied hillslopes, especially common in drylands on weak, easily erodible marl or shale bedrock, usually with little or no vegetation cover. They are usually characterised by a very high drainage density of rills and gullies. In some areas badland development is the result of intense human-induced soil erosion.

barchan [80]**:** a crescentic sand-dune form, with a smooth 'back' facing the prevailing wind, a steeper crescentic avalanche face facing downwind that culminates in two arms.

barrier reef [94]**:** an offshore coast-parallel coral reef, separated from the shore by a lagoonal area.

basalt [7, 52]**:** a common volcanic rock, dark in colour, usually formed as lava flows. A basic as opposed to an acid composition, dominated by ferromagnesian minerals (pyroxenes and olivine) together with some feldspar. May be characterised by columnar hexagonal joints, contraction cooling cracks.

base level [31]**:** the lower elevational limit of subaerial erosion processes. This may be local, as in the case of a resistant rock horizon or a main valley floor, or may be regional, as in the case of sea level. A base-level fall is a major cause of drainage incision.

baseflow [13]**:** as opposed to floodflow, baseflow is the dry-weather component of river flow, sustained by groundwater flow.

beachrock [92]**:** $CaCO_3$-cemented beach sands occurring within the intertidal zone or just above on tropical coasts.

bedload [66]**:** the component of a river's sediment load carried close to the bed by tractional processes or near the bed by saltation, distinct from the suspended load that is held in suspension in the water column. Bedload comprises the coarser component of the total load: sand, gravel, cobbles and boulders.

bedrock channel [67]**:** a river channel cut into bedrock, in which stream power is too high to allow the accumulation of sediment (in contrast with **alluvial channels**).

belt of no erosion [57]**:** the crestal zones on slopes undergoing gully erosion (**badland** areas), near to the divides, so that the run-off generated from upslope is insufficient to cause turbulence of the flow, and therefore is incapable of generating shear stresses sufficient to cause erosion. Processes other than erosion by overland flow therefore dominate these zones: surface cracking through, for example, wetting and drying, particle removal by wind, chemical processes.

boulder clay [85]**:** an old-fashioned term to describe poorly sorted glacial deposits ('till')comprising all size ranges from boulders to clay, usually with a bimodal distribution, which may now be referred to as 'dimicton'.

braided river channel [71]**:** a multi-thread river channel in which the channel divides around sand bars, gravel bars or vegetated islands. Characteristic of high energy bedload dominated rivers. Distinct in form and behaviour from lower energy, lower gradient, mud-dominated, much more stable, *anastomosing channels.*

breccia [102]**:** a sedimentary rock composed of angular stone-sized particles. *Brecciation* refers to the *in situ* fracturing of a massive rock into angular fragments by weathering processes. A common phenomenon in mature **calcretes**.

C

calcite [33, 52]**:** a mineral, $(CaCO_3)$, soluble in weak acids (rainwater, humic acids), the primary constituent of limestone, responsible for the dominance of solutional forms in limestone (karst) regions.

calcrete [55]**:** an indurated layer formed of calcite. Pedogenic calcrete forms by the downward movement of $CaCO_3$ and its precipitation some way below the surface as a Bk or K horizon in areas with a sustained soil moisture deficit, i.e. semi-arid regions. On exposure such horizons become indurated, undergoing a complex sequence of brecciation and recementation. They may form a duricrust or caprock. Groundwater calcrete forms in similar areas, but in zones of the soil profile just above the water table or where there is a reduction in vertical permeability. *See* also **duricrusts**.

canyon [67, 105]**:** incised river valley or gorge, characteristic of rapid incision in dryland regions.

caprock [33]**:** a resistant layer, which may be a resistant rock layer or a **duricrust**, protecting a hilltop or escarpment from erosion, underlain by a weaker layer more prone to erosion, the two giving a cliff face over a concave slope morphology.

case hardening [56]**:** the concentration of minerals in the outer layer of a rock by precipitation through desiccation. Protects the outer surface of the rock from erosion. May lead to the formation to **tafoni** (honeycomb weathering).

chelation [52]**:** a chemical weathering process involving the removal of metallic ions from clay minerals. Important in pedological processes but also in the weathering of clay-rich rocks.

chronosequence [100]**:** a set of related soils whose main differences are due to age differences: useful in estimating the relative ages of geomorphic surfaces.

cirque [15, 84]: a bowl-shaped depression carved by glacial ice at the head of a mountain glacier system.

clasts [56]: rock fragments.

colluvium [65]: ill-sorted fine sediment accumulating at the foot of hillslopes through diffuse soil erosion or creep processes.

complex response [7]: a term related to geomorphic systems, coined by Stanley Schumm, whereby the response to a change triggers the opposite trend. e.g. a change of flow regime that results in the incision of a hitherto braided reach of a stream into a single-thread channel that may release sufficient sediment to cause a hitherto single-thread reach downstream to switch to a braided regime.

conglomerate [104]: a sedimentary rock composed of rounded pebble- to cobble-sized *clasts*.

consequent drainage [38]: initial drainage pattern, created on a newly exposed or tectonically uplifted surface, in which the drainage direction follows the original slope.

conservative plate boundary [10]: boundary between two tectonic plates at which there is neither creation nor destruction of the crust. The two plates tend to move laterally against each other along a major fault or fault system, creating a major earthquake hazard (e.g. San Andreas fault, California; Anatolian fault, Turkey).

constant slope [65]: a straight slope, usually in mid-slope (at the **angle of rest** of loose material, or on bedrock at the **angle of incipient movement** of loose **clasts**.

constructive plate boundary [7]: boundary between two tectonic plates at which new oceanic crust is created by rifting and sea-floor spreading, coincident with mid-ocean ridges. Major zone of basaltic volcanic activity (e.g. the mid-Atlantic ridge).

continental crust [7, 9]: thicker, lighter crust, composed dominantly of silica-rich (granitic) rock, giving greater elevation to the continents than to the ocean basins.

continental shelf [16]: the edge of the continents, composed of continental crust, but flooded by ocean water to depths of *c.*200 m. Much of the continental shelf was exposed as land during the low eustatic sea levels during Pleistocene glaciations.

continental slope [16]: the edge of the continental shelf, sloping down to the abyssal plain.

Coupling/connectivity [p. 65]: The linkage between different components of the sediment cascade; eg. hillslope-to-channel coupling expresses the efficiency of sediment transfer from the hillslopes to the channel system, tributary-junction and reach-to-reach coupling expresses the sediment transport connectivity between different parts of the channel network.

coral reef [94]: coral is the $CaCO_3$ exoskeleton secreted by colonial marine 'polyps' to form coral rock. The polyps live in warm carbonate-rich tropical seas down to depths of tens of metres, and build the reef toward the surface. On exposure above sea level the coral dies, leaving coral rock. Reef forms include **fringing reefs** around the shore, **barrier reefs** somewhat offshore, and **atolls** – circular reefs around now submerged islands.

corestones [53]: a relatively unweathered boulder surrounded by *in situ* weathered material, a common feature produced by the chemical weathering of massive igneous rocks (e.g. granite).

cosmogenic dating [103]: rock exposed at the surface of the earth is bombarded by cosmic rays which interact with the minerals in the rock to produce a set of cosmogenically generated nuclides (especially ^{10}Be but others as well). These nuclides progressively accumulate within the rock decreasing in concentration with depth below the surface. Cosmogenic dating, a specialist laboratory procedure, measures the concentration of cosmogenically derived nuclides within rock profiles, so has the potential for dating rock exposure, provided that aspect and exposure history are talken into account.

creep [60]: *see* **soil creep**

cryoturbation [51]: disturbance of the soil or *regolith* by frost action, especially important during seasonal refreezing of the active layer above permafrost.

D

dead ice topography [85, 87]: the topography produced by a mass of glacial ice that has been cut off from its source and 'dies' *in situ*; the resulting terrain is a hummocky moraine, composed of ill-sorted, poorly consolidated sediments.

debris flow [62]: a mobile mass of unconsolidated rocks and fine sediments with variable water content moving downslope, often within a gully. Its source may be a slope failure, or may be simply within-gully sediment. Movement is usually triggered by heavy rains. Movement ceases as slopes decrease. On deposition may form a debris-flow lobe; important constituents of alluvial fans from small, steep catchments. Can be very hazardous to human life and settlements, especially in the case of **lahars**, debris flows triggered by volcanic activity.

dendritic drainage [38]: a randomly branching drainage network.

dendrochronology [102]: tree ring dating.

deranged drainage [29, 40]: irregular drainage patterns showing little or no organisation or relation to underlying structure. Characteristic of former glaciated depositional or erosional terrain.

desert pavement [51]: a stony desert surface

(also known as stone pavement, gibber – in Australia; reg – in the Middle East), produced by the mechanical weathering of bedrock or gravel surfaces, contemporaneously with the accumulation of windblown silts, which become trapped below the pavement layer.

Desert Varnish (or rock varnish) [102]**:** A coating of rock surfaces in desert environments, with iron and manganese oxides, giving dark brown to black colour to exposed surfaces (Fe and Mn compounds) and a red colour (Fe compounds) to the undersides of stones (eg. on desert pavements). Thought to be produced by bac terialogical activity. Degree of development increases (varnish darkens) with age of exposure.

destructive plate boundary [8]**:** boundary between two tectonic plates where crust is consumed by subduction.

doline [35]**:** a small, generally circular depression produced by sub-surface solution of limestone bedrock in *karst* areas. May provide a routeway for surface streams to pass underground (a swallow hole).

dip [36]**:** the angle of inclination of sedimentary strata: Uniclinal dip relates to dip at a constant angle in a constant direction.

drainage density [60]**:** the length of stream channels in relation to drainage area, usually expressed in km/km^2. Drainage densities are very high in **badland** areas and very low in **karst** areas.

drumlin [87]**:** a smooth, elongate low hill produced by subglacial deposition and smoothed in the direction of ice movement.

duricrust [55]**:** a resistant caprock produced by the exposure of horizons of accumulation of salts within the soil profile. Includes 'calcrete' ($CaCO_3$) accumulations in semi-arid to arid zones, 'gypcrete' (gypsum anhydrite – $CaSO_4$. n H_2O) in arid and hyper-arid zones, silcrete (SiO_2) and ferricrete, formerly known as laterite (rich in Fe_2O_3), in the seasonally wet/dry tropical zone.

dynamic equilibrium [14]**:** a form of equilibrium whereby a persistent morphology is maintained by a balance between input (e.g. deposition) and output (e.g. erosion).

E

epeirogenic (post-orogenic) uplift [29, 45]**:** regional tectonic uplift of a zone of crustal thickening produced by past plate-tectonic activity in a mountain belt (e.g. after subduction ceased in the European Alps in the Miocene the whole region has been uplifted during the Pliocene and Pleistocene). Additionally, upwelling mantle processes may cause regional uplift (e.g. as in the Colorado Plateau, USA).

erosion surface [44]**:** this term has two meanings: in a sedimentological sense – a relatively small-scale feature, an erosional horizon cross-cutting the sedimentary structures of an underlying deposit. At a larger scale – extensive surfaces that cross-cut the underlying geological structure, often forming upland plateaux, characteristic of the 'older' mountain belts, the 'uplands' of western Europe. There is controversy concerning their origin, whether they are uplifted former **peneplains**, **pediplains**, marine-cut surfaces or **etchplains**.

esker [88]**:** an irregular linear ridge of sand and gravel in glacial depositional terrain, preserving the course of a former subglacial stream.

etchplain [44]**:** the lowering of the land surface by tropical deep-weathering processes to produce an extensive plain or plateau surface. It has been argued, especially by German geomorphologists, that this is the main mechanism that produced the extensive so-called erosion surfaces in western Europe during the late Tertiary.

eustasy [26]**:** as in *eustatic sea-level change* – variations in global sea level resulting from the different proportions of the world's water stored as glacial ice. Eustatic sea levels are relatively high during global interglacials (as now) and low during global glacials.

evaporite [56]**:** a chemically precipitated sediment produce by evaporation to dryness of a saline water body, e.g. of an arid-region playa lake. Most commonly composed of (in increasing order of solubility) calcium carbonate, gypsum, or halite.

evapotranspiration [11]**:** the sum total of moisture returned to the atmosphere by evaporation from open water and moist surfaces, and that transpired by the vegetation.

F

feedback (positive, negative) [15]**:** internal system mechanisms that either dampen the effects of an externally-induced change (negative feedback), or magnify and reinforce the effects of such a change (positive feedback).

ferrirete [55, 56]**:** *see* **duricrusts**.

fetch [88]**:** the extent of open sea facing a coast. The greater the fetch the greater potential for high wave energy.

field capacity [13]**:** the moisture content that can be retained within a soil by capillary forces against gravity. Unless the soil moisture has reached field capacity, downward percolation cannot take place. Soil moisture content in humid areas tends to be in excess of field capacity, whereas in semi-arid and arid areas it may rarely reach field capacity (soil moisture deficit).

firn [12]**:** the intermediate stage between snow and glacial ice. Essentially it is glacial ice that is white in colour due to the presence of air bubbles. On further compaction and refreezing it will be transformed into true glacial ice.

floodflow (quickflow) [12]**:** the component of river flow (as opposed to the **baseflow** component) generated by rapid surface run-off (overland flow), or rapid interflow of rapid snowmelt. Forming a flood peak in the **hydrograph**, and responsible for the majority of in-channel fluvial processes (bank erosion, sediment transport, especially of coarse sediments, deposition of gravel bedforms, etc.).

floodplain [74]**:** a flat plain adjacent to an alluvial channel, composed of alluvial sediments, and whose surface defines the limits of the alluvial channel.

free face [57]**:** an exposed rock face (a cliff), usually high on a slope, on which mechanical weathering and rockfall are the dominant processes.

freeze-thaw weathering [20]**:** the mechanical weathering of rock by ice forming within cracks or along bedding planes in the rock. On freezing, water expands as it turns into ice, thus creating stresses that weaken the rock, eventually fracturing it and yielding angular clasts.

Fringing reef [94]**:** *see* **coral reef.**

G

geomorphic sensitivity [7]**:** the relationship between the frequency of an erosional or depositional event and the recovery time of the land surface involved. For example, a sensitive landscape is one where recovery time, perhaps through re-vegetation, is slow in relation to the frequency of disturbing events. A robust landscape is one where recovery is fast in relation to the frequency of disturbing events.

geomorphic threshold [7]**:** a change in the erosional or depositional status of a landform or landscape, or a change in morphology, that affects the rate or direction of processes or the morphology itself (e.g. the switch from a non-gullied to a gullied hillslope, or the switch from a meandering to a braided river channel). Such thresholds may be brought about 'intrinsically' through mechanisms internal to the system (e.g. continued deposition may increase local channel gradients, so that the next flood event triggers erosion), or 'extrinsically' by an environmental change, such as a vegetation change, a climatic change or tectonic activity.

granite [52]**:** a coarse-grained, massive, crystalline igneous rock, composed dominantly of quartz, orthoclase (potassium-rich) feldspars, possibly also plagioclase (sodium-rich) feldspar, muscovite (white) mica, and possibly also biotite (dark) mica. Forms at depth as a **plutonic rock** in large igneous intrusions.

groundwater [13]**:** water stored at some depth below the water table (as opposed to soil moisture, stored within the soil). Recharged by percolation, discharged as springs.

grus [53]**:** products of the chemical weathering of granite, composed of little-altered quartz grains, mica flakes and abundant clay minerals.

gully [59]**:** two forms: (i) hillslope gully: eroding channel cut into a hillslope (ii) linear incised channel on a valley floor (*see* **arroyo**).

gully erosion [59]**:** downslope culmination of erosion by overland flow, water and sediment normally fed from rilled sideslopes. Extensive gully erosion normally termed '**badland**' erosion.

gypcrete: *see* **duricrusts.**

H

hanging valley [87]**:** a tributary valley whose valley floor is 'perched' above that of the main valley, usually as the result of glacial over-deepening of the main valley.

head deposits [23, 61]**:** periglacial hillslope deposits, comprising angular, freshly weathered clasts set in a fine matrix, fed downslope by solifluction processes.

Hjulstrom curve [66]**:** relationships derived by Hjulstrom in the 1930s between velocity of flowing water, particle size and entrainment, transport or deposition of sediment.

Holocene [5, 78]**:** the most recent period of geological time, roughly the last 10,000 years; Oxygen Isotope Stage 1, loosely the 'post-glacial' period.

hydration/dehydration [51]**:** weathering process involving the absorption into/loss of water from the chemistry of a material.

hydraulic geometry [15, 66]**:** relationships, first defined by Leopold and Maddock in the 1950s, between river discharge, width, depth, velocity and other hydraulic and sediment transport variables in river channels.

hydrograph [12]**:** graphical plot of river discharge against time.

hydrolysis [51]**:** chemical weathering process involving the exchange of ions: significant in the weathering of the minerals in igneous rocks.

I J

ice-wedge (casts) [51]**:** see *patterned ground.*

igneous rocks [32, 52]**:** rocks formed by crystallisation from magma, either within the crust as intrusive rocks or at the surface as volcanic rocks.

imbricate fabric [69]**:** when clasts (stones) are rolled along a stream bed by traction as bedload, they come to rest by friction with the bed, presenting a smooth face upstream into the current. This usually means that their long axes lie across the current, but their major planes (defined by their 2nd and 3rd axes) dip into the current, i.e. upstream. A bed of gravel deposited in this way shows an overlapping stacked fabric, not unlike that of roof tiles, with the dip of the clasts upstream, at an angle to that of the base of the bed. Bi-directional flows, such as those on shingle beaches, tend to produce a clast fabric where the dip of the clasts is more nearly parallel to that of the base of the bed.

Incised meanders [105]**:** Meandering valleys cut into

bedrock. Response to incision often generated by **epeirogenic** uplift.

infiltration capacity [11]**:** the rate at which the soil can absorb water by infiltration: varies with soil moisture, soil particle-size and pore space. Overland flow (surface run-off) can only occur if rainfall intensity exceeds infiltration capacity.

interception [11]**:** the trapping of rain or snow by the vegetation cover.

interflow [13]**:** the lateral drainage of excess soil moisture (i.e. in excess of the soil's field capacity). Rapid interflow can contribute to the floodflow component of streamflow.

inverted relief [37]**:** relief where high topography coincides with is structural lows and vice versa, e.g. synclinal ridges or anticlinal valleys.

involution [51]**:** contorted structures in the soil or regolith produced by **cryoturbation**. Produced during seasonal refreezing of the active layer over permafrost.

isostasy [7, 28]**:** vertical change in elevation of the crust, due to its buoyancy. Two forms are important: crustal and glacial isostasy. In crustal isostasy, zones of crustal thickening due to plate-tectonic activity are elevated, creating high mountain chains, or upland areas. Erosional offloading creates further epeirogenic uplift. In glacial isostasy, loading by glacial ice depresses the crust, which on deglaciation responds by rebounding. Similar effects may be produced by loading from water bodies.

isostatic sea-level change [26]**:** change in relative sea level due to isostatic change in the level of the land surface; usually used in the context of relative sea-level fall due to post-glacial isostatic rebound of the land.

K

kame [87]**:** a mound of fluvioglacial sand and gravel deposited by **dead ice** as the ice collapses on melting.

kame terrace [87]**:** a body of sand and gravel deposited by glacial meltwater in the space between decaying ice and the valley margin. A feature characteristic of **dead ice** decay.

karst [33]**:** a term coined from the 'Karst' region of former Yugoslavia to describe the geomorphology of solutional terrain, particularly limestone.

kettle hole [87]**:** a depression caused by the melting of a block of ice, detached from the main body of ice during deglaciation. May often contain a small circular lake, a kettle lake.

knickpoint [68]**:** a step or a break in the longitudinal profile of a stream caused by incision working its way upstream.

kopje [53]**:** an African term for a **tor**, a bouldery isolated hill.

L

Lahar [63]**:** a debris flow fed or augmented by volcanic eruption.

laterite [56]**:** *see* **duricrust**.

lichenometry [102]**:** the estimation of the age of a deposit (e.g. a moraine or a fluvial gravel terrace) or the age of rock exposure from measuring the size of lichens that have colonised the stones or the rock surface, and applying (usually) known lichen growth rates.

limestone pavement [33–5]**:** bare limestone rock surfaces, usually coincident with the bedding planes, and from which the overlying rocks have been stripped (usually) by glacial action, and on which solutional features are apparent.

liquid limits [60]**:** the moisture content at which a fine-grained (usually clayey) sediment or soil drains, and flows as a liquid.

lodgement till [85]**:** **boulder clay** deposited under compression at the base of a glacier or ice sheet, so that it has a compact texture. It may show clast alignment with the direction of ice flow.

loess [26, 79]**:** windblown silts that accumulated in sheets around the margins of continental-scale Pleistocene ice sheets.

longshore drift [91]**:** movement of sediment along the beach, produced by oblique wave approach, often producing shore-parallel depositional features such as spits.

luminescence dating [103]**:** quartz or feldspar crystals within buried sediments absorb background ionising radiation, which can be released with stimulation in specially equipped laboratory settings as luminescence. The luminescence is bleached away by exposure to sunlight. The luminescence signal, which increases with burial time, can be used to date the last exposure of the sediment to sunlight. This method has been successfully applied to dating quartz- or feldspar-rich sandy sediments, particularly from **aeolian** or **fluvial** environments, environments where exposure to light occurs during transport, followed by rapid burial on deposition. It is applicable for late Quaternary timescales of the last 100,000 years or so. In contrast with most of the radiometric decay methods available over these timescales, it gives a date of sedimentation, rather than a before or after date.

M

magnitude and frequency concept [6]**:** the relationships between event frequency and magnitude (expressing the total geomorphic work done) – a concept first introduced by Wolman and Miller in the 1960s.

mangrove [94]**:** tropical salt-tolerant shrubs that root in seawater in low energy tropical coastal environments. Once established, they act to trap further fine sediments.

mass balance (of a glacier) [84]**:** the relationship between snow accumulation and glacial ice *ablation*. Controls the rate of ice movement and the position of the ice front.

meandering channel [71]**:** single-thread alluvial channels, whose plan view comprises a series of bends alternating from side to side.

meltout till [85]**:** a glacial deposit composed of loose, unsorted material (generally boulder clay) that has been deposited in situ by melting of the supporting ice.

metamorphic rocks [32, 52]**:** rocks (originally igneous or sedimentary) whose properties (physical and/or chemical) have been radically altered by heat and/or pressure.

mid-ocean ridge [7]**:** *see* **constructive plate boundary**.

Milankovitch cycles [4]**:** cyclic variations in the Earth's orbital characteristics, including a 96,000 year eccentricity cycle, a 40,000 year obliquity cycle and a 21,000 year precessional cycle, described by Milankovitch in the 1920s. Though Quaternary climatic fluctuations can be shown to bear some relationship with the combined effects of these cycles, the exact mechanism whereby these cycles might influence global climates is still uncertain.

misfit (underfit) stream [68]**:** (generally) a modern meandering alluvial river channel set in a much larger meandering valley, the implication being that as the geometry of the modern meanders is adjusted to modern flow conditions, the much larger valley meanders would be adjusted to much larger discharges in the past. There is a flaw to this argument in that the modern meanders are mobile meanders set in an alluvial floodplain and free to adjust by both erosion and deposition, whereas valley meanders are erosional forms, cut in bedrock and unable to adjust. They represent the cumulative effects of (admittedly) large flow events.

moraine [84, 87]**:** two meanings – (i) the sediment carried by a glacier on its surface (supraglacial moraine), internally (englacial moraine), marginally (lateral moraine), (becoming, where two glaciers join, a medial moraine) or at the base of the glacier (subglacial moraine); and (ii) the features produced by the deposition of moraine – at the snout of a glacier (terminal moraine), at the limit of an ice sheet (end moraine), at stillstands during glacial retreat (recessional moraines), and at the sides of a glacier (lateral moraine).

morphometry [42]**:** measurements of the relief, areal extent, size, shape, slope angle or other geometric properties of landforms (e.g. slopes, drainage basins, river channels, alluvial fans, glacial features) and the numerical relationships between them.

moulins [85]**:** circular vertical funnels in the surface of a glacier, down which supraglacial meltwater may pass into the glacier.

N

nappe [37]**:** a forward-thrusted overfold produced during mountain building on a destructive plate boundary, characteristic of complex mountain chains such as the Alps.

neotectonic [45]**:** ongoing tectonic activity, continuing after the main mountain-building phase. In 'Alpine' young fold mountains, normally implies tectonics continuing into the Neogene.

nivation [20]**:** mechanical weathering by freeze-thaw in relation to a snowpatch.

O

oceanic crust [7]**:** crust that is thinner and denser than continental crust, composed primarily of ferromagnesian-rich (basaltic) rocks.

order, as in stream order [42]**:** a system first devised by Horton, modified by Strahler, of classifying stream segments according to their position in the hierarchy of branching, so that unbranched headwater streams are 1st order streams, two 1st order streams join to form a 2nd order segment, two 2nd order segments to form a 3rd order segment, and so on.

overland flow [12, 59]**:** surface run-off generated by rainfall excess or saturation overland flow.

oxidation/reduction [51]**:** chemical processes important in the weathering of rocks and soil formation by the addition/loss of oxygen. Particularly important in iron-rich environments, involving transformations between ferric and ferrous iron compounds.

P

palaeosol [26]**:** a soil formed in the past and no longer undergoing the processes that formed it. This may be a relict soil on the surface or a 'fossil' soil below a layer of younger deposits. Palaeosols may be useful indicators of past environments, and can often be used in the relative dating of land surfaces.

paraglacial [25]**:** the period of heightened geomorphic activity following deglaciation. Usually involves the reworking of sediment released by deglaciation.

patterned ground [51]**:** two forms, both characteristic of permafrost environments: (i) polygons formed by ice wedges – in section, may be expressed as ice-wedge casts; and (ii) sorting of stones into polygons, garlands, or stripes (depending on slope steepness), during active-layer seasonal refreezing as the result of the different thermal properties of stones and matrix – in section, may be expressed as **involutions**.

pediment [59, 65]**:** rock surface formed by the parallel retreat of slopes, common in dry regions.

pediplain [44]**:** extensive coalescent pediments: forming an erosion surface produced by the parallel retreat of hillslopes/escarpments.

pedogenic [55]**:** relating to soil processes.

peneplain [43]**:** supposed end product of the 'cycle of erosion', produced by the 'downwaring' of slopes close to a stable base level – nowadays a disputed term.

periglacial [11, 23]**:** the area surrounding continental-scale ice sheets. When applied to processes, the presence of permafrost is implicit.

permafrost [20, 23, 62]**:** permanently frozen ground at depth. The surface layer (the active layer) thaws during summer and refreezes in autumn. Stresses caused by refreezing from the surface down are responsible for many features characteristic of permafrost (periglacial) environments (such as patterned ground, involutions, ice wedges)

phytokarst [93]**: karstic** dissolution accelerated by acid-secreting algae – common on tropical or sub-tropical limestone coasts or on exposed coral rock on such coasts.

pipe erosion [59]**:** subsurface erosion of tunnels, especially important in badland areas.

plastic limits [60]**:** the moisture content at which a fine-grained (usually clayey) sediment or soil deforms under its own weight by plastic flowage.

plate boundary [7]**:** see **constructive**, **conservative**, **destructive plate boundaries**.

playa [56]**:** ephemeral desert lake, normally dry and forming a 'salt pan'.

plutonic rocks [49]**: igneous rocks** that crystallised slowly in large, deep-seated igneous intrusions.

podzol [55, 100]**:** characteristic acid soil of cool temperate environments, showing well-developed horizon – from the surface down a dark humic horizon, a pale bleached horizon from which the iron compounds have been leached into a reddish broom **illuvial** horizon below.

polje [35]**:** a medium to large solutional or collapse depression in **karst** areas.

pools and riffles [70–71]**:** alternating deeps and shallows along (especially) a single-thread alluvial stream channel, their spacing dependent on channel width. Sustained by three-dimensional (secondary) flow cells, which themselves are instrumental in **meander** formation from a pool and riffle sequence.

pressure melting point (of glacial ice) [83]**:** the combination of pressure and temperature at which glacial ice melts – important subglacially in differentiating cold-based from temperate-based glaciers.

pressure release (offloading) joints [49]**:** a major form of mechanical weathering. Surface-parallel joints produced by the elastic response of rock to either erosional offloading or offloading related to the melting of an overlying thickness of glacial ice.

proglacial [85]**:** the zone in front of a glacier or ice sheet.

prograde/progradation [75]**:** an extension by deposition of a depositional landform in the direction of flow (e.g. a delta may prograde into the sea), or an alluvial fan may extend its distal limits by deposition.

Q R

quartz [52]**:** silicon dioxide (SiO_2) – a common rock-forming mineral, abundant in acid igneous rocks. Physically strong and almost chemically inert, therefore persists through weathering cycles to be abundant in sedimentary rocks. The primary constituent of sand, and therefore of sandstone.

Radiocarbon dating [102]**:** The cosmic ray flux into the upper atmosphere produces radioactive ^{14}C atoms which are rapidly oxidised to $^{14}CO2$ and become thoroughly mixed with the existing atmospheric $^{12}CO_2$, and incorporated into livig tissue. When atmospheric input ceases, on death of an organism, the radioactive ^{14}C atoms decay to ^{12}C, emitting Beta particles as they do. With time the tiny proportion of ^{14}C atoms decreases, as does the activity level. Standard radiocarbon dating measures the activity generated by the release of Beta particles and calibrates this with the known curve for radioactive decay of ^{14}C. Conventional radiocarbon dating requires relatively large samples of organic matter, and can be used for estimating ages back to about 30,000 years. The AMS method uses a different approach. After pre-treatment the sample is accelerated in a particle accelerator, allowing individual atoms to be counted on the basis of their mass. The proportion of ^{14}C atoms can be identified and a radiocarbon date derived. Much smaller samples can be dated than by the conventional method, and the reliable dating timescale can be extended considerably.

raised beach [95]**:** former beach deposits, related to a previous higher sea level and preserved above modern sea levels.

recurrence interval (return period) [6]**:** the average (or most probable) interval between events of a similar magnitude (floods, earthquakes).

reef [94]**:** *see* **coral reef**.

regolith [49, 60]**:** superficial layer of weathered bedrock and soil.

rejuvenation [31, 38]**:** the response of a river system to uplift or a fall in base level, involving incision and the steepening of river profiles.

relief, available relief [1]**:** height range within which geomorphic processes may operate; lower limit defines by local base level.

resequent relief [37]**:** relief where high topography coincides with structural highs and vice versa, e.g. synclinal valleys or anticlinal ridges.

rill erosion [59]**:** erosion on hillslopes by overland flow, where sheetflow begins to winnow out shallow channels that have more or less the same gradient as the slope itself. These channels may be ephemeral. These channels converge downslope into gullies, channels that are capable of eroding the underlying substrate and have a gradient markedly less than the slope gradient.

river capture [39, 105]**:** aggressive drainage cuts back and intercepts an earlier line of drainage, diverting the drainage into the new course. The 'beheaded' stream loses its headwaters. The captor stream now becomes the major stream. Upstream the whole system may undergo rejuvenation and incision, due to a lower local base level.

roche moutonnée [85]: rock outcrop smoothed by glacial erosion. The up-ice face is smoothed, the down-ice face is usually irregular, due to 'plucking' by the overriding ice.

run-off [12]: loosely, the catchment water yield or average river flow; more specifically, overland flow.

S

sabkha [80]: an Arabic term relating to saline deposits in arid environments, including coastal sabkha accumulating in coastal lagoons, and inland sabkha, accumulating in depressions (e.g. interdune areas) inland.

saltation [80]: an intermediate process in sediment transport between particles transported along the bed (of a stream in fluvial processes) or surface (for sand particles, transported by the wind) and those in suspension in the fluid (water in the case of fluvial processes, air in aeolian processes). In both cases particles tend to be sand-sized. Individual sand particles move by a series of 'leaps', never far from the bed or surface, and never in true suspension. Under the very high energy conditions of a large, steep river in flood, larger particles may move by saltation.

salt marsh [93]: a low-energy coastal zone, either within an estuary or protected from wave action by a spit, within which mud is deposited to form mud flats, which are colonised by salt-tolerant vegetation. The vegetation in turn fosters further sedimentation.

salt weathering [93]: mechanical weathering of rock or stones in arid or coastal areas induced by salt crystallisation in cracks or pore spaces in the rock or stone.

sandur [85, 88]: an Icelandic term meaning a sheet of fluvioglacial outwash gravels deposited by a proglacial stream, and forming a gravel plain interlaced by distributary braided channels.

scree (talus) [56]: a depositional slope formed by angular clasts derived from mechanical weathering of exposed rock above. The scree slope accumulates at the angle of rest of the particles.

sediment cascade (including sediment transport) [10, 13]: the sequence of events and the pathways taken from weathering of bedrock, through its erosion, transport and deposition through the range of geomorphic systems.

sedimentary rocks [32, 52]: rocks formed by the accumulation and lithification of materials (sediments) that have passed through a cycle of weathering, erosion, transport and deposition. Includes clastic sedimentary rocks (e.g. sandstones), composed of fragments, and chemical precipitates.

sensitivity (geomorphic sensitivity) [7]: the relationship between the frequency of threshold-exceeding events and the time taken for the system to recover from threshold exceedence (recovery time). A 'sensitive' landscape takes a long time to recover, whereas a 'robust' landscape recovers rapidly.

sheet erosion [59]: slope erosion by unconfined overland flow, that is capable of winnowing loose material from the surface. Winnowing may result in the flow becoming concentrated into rills.

shield [19]: 'core' area of a continent, formed of ancient (Precambrian) usually metamorphic rocks. These tectonically relatively stable areas have remained far from plate boundaries since throughout the Phanerozoic (e.g. the Canadian Shield, the Baltic Shield).

silcrete [55]: see *duricrust*.

sinuosity [71]: in a meandering river channel the ratio between channel distance and the straight line distance between two points along the channel. Rather arbitrarily, channels with a sinuosity > 1.5 are deemed to be meandering.

soil creep [60]: the slow (imperceptible) downslope movement of soils as the result of a downslope gravitational component affecting swelling and shrinking of the soil in response to wetting and drying (also to freezing and thawing).

solifluction [20, 62]: term loosely used to define the unconfined mass movement of unconsolidated material downslope by flowage processes, more strictly related to downslope movement of the active layer in periglacial or permafrost environments.

solution (of limestone, etc.) [33, 51]: the ionisation of soluble materials and the removal of the separate ions in solution. Limestone is soluble in weak acids, such as rainwater and humic acids. Limestone ($CaCO_3$), gypsum ($CaSO_4$ n H_2O) and halite (NaCl) are common soluble rocks or minerals (listed by increasing solubility).

stream power [66]: an expression of the energy available for erosion and sediment transport in rivers and streams. Total power (expressed in watts) increases with discharge and gradient. Unit power, the power available at a point in the channel, increases with depth and gradient. The threshold of critical power, outlined by Bull in the 1980s, expresses the relationship between critical power (the power to transport the sediment supplied) and the actual (unit) power.

subduction zone [9]: the zone, occurring at a destructive plate boundary, where oceanic crust is forced below the continental crust and is eventually re-absorbed into the mantle.

subsequent drainage [39]: drainage directions that exploit lines of weakness in the underlying rock. They modify the initial (consequent) drainage directions, primarily by river capture.

superimposed drainage [40]**:** a river system initiated as consequent drainage, which incised through the surface layers of rock and through an unconformity into an older, different set of rocks below the unconformity. The river now runs discordantly with the structures of the underlying rocks.

Supra-glacial moraine [84]**:** *see* **moraine**.

suspended sediment load [66]**:** fine-grained sediments that are transported in suspension in the water column as opposed to bedload, which is transported near to the river bed.

syncline [37]**:** a downfold in bedded rocks in the form of a saucer. If the fold preserves a ridge of the upper (younger) rocks in the centre it will be defined as a synclinal ridge, a case of inverted relief.

T

tafoni [56]**:** honeycomb weathering whereby the surface of the rock is more resistant (case-hardened) than inside, which is then easily eroded.

talus [56]**:** see *scree*.

tectonics [4]**:** crustal processes involving deformation of rock bodies and uplift.

terrace; river terrace [44, 78]**:** former floodplain surface, into which the river channel has since incised. Forms a flat surface well above the modern flood level of the river. A series of terraces may form a staircase of flat surfaces.

terracettes [61]**:** small-scale (up to about 25 cm high) features characteristic of steep, grass-covered hillslopes, which form small steps subparallel with the contour of the hillslope. Sometimes erroneously described as 'sheeptracks'. They are probably formed by a combination of creep and micro-slip of the turf layer and the soil immediately underneath.

threshold (geomorphic threshold, intrinsic, extrinsic thresholds) [7]**:** a radical change of state of the geomorphic system or of part of it (e.g. the switch from a single thread **meandering** channel to a multi-thread **braided** channel; the switch from an ungullied hillslope to a gullied hillslope; the switch from an aggrading or stable stream channel to an incising channel). Such a change may be induced by the internal characteristics of the system, particularly through the operation of positive **feedback** mechanisms (an intrinsic threshold), or by a change in the environmental controls (an extrinsic threshold: e.g. a tectonic, a climatic or a land-use change).

till (glacial till) [85]**:** poorly sorted glacially deposited sediment comprising coarse clasts in a silty or clayey matrix (also described as **boulder clay**; may now be referred to as 'dimicton'). Includes *lodgement* till, deposited subglacially, and *meltout* till, deposited at the glacier margins.

tor [53]**:** an upstanding exposure of rock on a hilltop area or on a valley side. Usually produced by a combination of mechanical (**pressure**

release jointing, may be modified by freeze-thaw weathering) and chemical weathering (whereby the surrounding material has rotted and been removed by erosion, leaving **corestones** that form the basis of the tor). Common in granite terrain, but do occur in other lithologies. In the tropics (especially in Africa) similar features are described as **kopjes**.

transform fault [10]**:** a major lateral fault either within a tectonic plate (accommodating different rates of plate movement) or between two tectonic plates (at **conservative plate boundaries**).

trellis(ed) drainage [39]**:** a rectilinear drainage pattern produced by the development of **subsequent drainage** along parallel outcrops of weaker rocks.

U

unconformity [40]**:** an erosion surface that truncates one set of geological structures, buried by a younger, less deformed set of rocks.

uniclinal [36]**:** as in uniclinal dip of a geological bed or layer of rock, dipping consistently in one direction (in other words, dipping but not folded – at the scale of what is being described).

uvula [35]**:** a large solutional or collapse depression in karst areas.#

V

Vauclusian spring [35]**:** resurgence of an underground river in karst areas. Named after the Vaucluse area in southern France, a limestone plateau around whose margins several underground rivers emerge.

W

wave-cut platform (rock platform) [90]**:** a subhorizontal coastal erosional surface, sloping gently seawards.

weathering [10, 20, 49–56]**:** the mechanical and chemical breakdown of rock, yielding rock fragments (clasts), chemically altered rock, and ions that are removed in solution. A prerequisite for the operation of the sediment cascade.

X Y Z

yardang [79]**:** a landform created by wind erosion in arid regions of bedrock or of cemented surficial sediments. May have an irregular, semi-streamlined morphology.